WORD PROBLEMS

with Answers

Sam is 20 years older than Amy. 8 years ago, Sam was twice as old as Amy. How old are Sam and Amy now?

	8 years ago	now
Amy	20	28
Sam	40	48

Chris McMullen, Ph.D.

Word Problems with Answers

Chris McMullen, Ph.D.

www.improveyourmathfluency.com

www.monkeyphysicsblog.wordpress.com

www.chrismcmullen.com

Zishka Publishing

ISBN: 978-1-941691-54-0

Mathematics > Word Problems

Mathematics > Arithmetic

CONTENTS

EXAMPLES, TIPS & MORE

Are you looking for examples or problem-solving tips?

You can find examples and tips in the Idea Center on page 105.

Are you looking for specific types of problems?

You can find an index of problems sorted by topic on page 189.

Would you like to check your answers?

You can find an answer key at the back of the book.

EXERCISES

1. Determine the ratio of vowels to consonants for this sentence.

2. In a game that kids made up, one elephant is worth 15 giraffes, one giraffe is worth 8 zebras, and one zebra is worth 6 aardvarks. How many aardvarks are 3 elephants worth?

3. A student read pages 17-22, 39-51, and 74-75 in a textbook. How many pages did the student read all together? Note: There is a "trick" to this question.

4. Two numbers have an average value of 139 and a difference of 84. What are the numbers?

5. Fred is three times as old as Bonnie. The sum of their ages in years is 36. How old is each person?

6. It is presently 1:08 p.m. Quentin has been driving for 75 minutes. What was the time when Quentin began driving?

7. A green rectangle has a height of three inches. An orange rectangle has twice the length of the green rectangle and six times the area of the green rectangle. What is the height of the orange rectangle?

8. Three out of every five students are wearing jackets today. If 45 students are wearing jackets, how many students are **not** wearing jackets?

9. If a butterfly flaps its wings 480 times per minute, how many times does the butterfly flap its wings in 12 seconds?

10. A boy travels with a constant speed of 5 m/s to the east for 24 seconds. The boy then travels 80 m to the north in 16 seconds. The boy then travels 120 m to the west with a constant speed of 3 m/s. The boy made one more trip, returning to his starting point two minutes after he began. How fast and in which direction did the boy travel in the trip that returned him to his starting point?

11. Three girls earned a total of $41 for doing yardwork. When the girls split the money equally among themselves, they discover that there is a remainder. How much money does each girl receive? What is the remainder?

12. Since last year, the value of a car diminished by $2700, corresponding to a 15% drop in value. What is the value of the car now?

13. Theresa has $10. Pencils cost 72 cents each. Erasers cost 14 cents each. There is no sales tax. Theresa buys as many pencils as she can afford. With the money that is left over, Theresa buys as many erasers as she can. How many pencils and how many erasers does Theresa buy? How much money will Theresa have left?

14. Eight people attend a meeting. When the meeting ends, every person shakes hands once with every other person. What is the total number of handshakes?

15. A pendulum takes 2.5 seconds to oscillate back and forth. How many times will the pendulum oscillate back and forth in one minute?

16. Two numbers have a sum of 67 and a difference of 29. What are the numbers?

17. Presently, Victor is three times as old as Maria. Ten years ago, Victor was seven times as old as Maria. How old is each person now?

18. Iris only has quarters and Oscar only has dimes. Oscar has three times as many dimes as Iris has quarters. The value of Oscar's money is $2.25 more than the value of Iris's money. How many coins does each person have?

19. A rectangle measures 8×10. A man wishes to cover the rectangle with tiles that each measure 2×2. All measurements are in the same units. How many tiles does the man need?

20. How many different 5-character codes can be made using two J's, one Q, one 4, and one 8? An example of a valid code is 4JQ8J.

21. A wooden math puzzle costs $16 before sales tax and $17.36 after sales tax. What is the percent tax rate?

22. Initially, Natalie had 36 more binder clips than Claire. After Natalie gave one-third of her binder clips to Claire, Natalie had 16 fewer binder clips than Claire. How many binder clips did each girl have before and after?

23. Along a hallway, rooms are numbered counting by 4's from 140 thru 172. How many rooms are there?

24. A particular car can travel 45 miles along the highway on one gallon of gas. How far can the car travel along the highway on 14 gallons of gas?

25. In an experiment, when the pressure is 3 Pascals, the temperature is 240 Kelvin. When the pressure is 5 Pascals, the temperature is 400 Kelvin. When the pressure is 8 Pascals, the temperature is 640 Kelvin. Predict what the temperature will be when the pressure is 17 Pascals.

26. The sum of three consecutive odd numbers is five less than five hundred. What are the numbers?

27. A boy has two unusual dice. The unusual feature is that the sides are all odd numbers: 1, 3, 5, 7, 9, and 11. If the two dice are fair, what is the likelihood that the numbers will have a sum equal to 10 when they are rolled?

28. In six years, Wendy will be twice as old as she had been eight years ago. How old is Wendy now?

29. What is 15% of 30% of 60?

30. A classroom has 141 calculators. A cabinet has several identical drawers. Each draw can hold 18 calculators. A teacher fills as many drawers as possible with the calculators and places the remaining calculators in the top drawer. How many calculators are in the top drawer?

31. A robotic cat and robotic mouse are initially at rest next to one another. The robotic mouse begins to move with a constant speed of 8 m/s. Six seconds later, the robotic cat pursues the robotic mouse with a constant speed of 10 m/s. How much time will it take for the robotic cat to catch the robotic mouse?

32. A store sells 8 markers for $3.50. There is no sales tax. How much would it cost to purchase 56 markers?

33. A map is scaled such that 0.5 inches corresponds to 20 miles. On the map, two cities are 3.25 inches apart. What is the actual distance between the cities?

34. A list contains 60 numbers. Each number is either even or odd. There are 12 more odd numbers than even numbers. What is the ratio of the number of odd numbers to the number of even numbers?

35. On a hot day, a metal rod is 2% longer than it is at room temperature. A wooden meterstick is not noticeably longer than it is at room temperature. The hot metal rod is 76.5 centimeters long according to the meterstick. How long is the metal rod at room temperature?

36. On a hot day, a metal meterstick is 2% longer than it is at room temperature. A wooden rod is not noticeably longer than it is at room temperature. According to the hot metal meterstick, the wooden rod appears to be 76.5 centimeters long. What is the actual length of the wooden rod?

37. Four students are running a relay race. The first student runs a lap in 11.2 seconds. The second student runs a lap in 10.6 seconds. The third student runs a lap in 10.5 seconds. The school record is 42.1 seconds. How much time does the fourth student have in order for the team to match the school record?

38. An order was placed for a product on October 29. The receipt states that the product will arrive in 5 to 10 weeks. What are the earliest and latest dates for which the product should arrive?

39. Raj has the same number of quarters, dimes, nickels, and pennies. The total value of Raj's coins is $6.15. What is the total number of coins that Raj has?

40. Each pen in a drawer either has black ink or blue ink. For every 5 pens that have blank ink, there are 4 pens that have blue ink. There are 180 pens that have black ink. How many pens have blue ink?

41. The drama club sold 60 children's tickets for $2.70 each and sold 30 adult tickets for $3.60 each. What was the average cost of a ticket?

42. Tyrone is 29 years old. Pete is 53 years old. When was Pete three times as old as Tyrone?

43. In this problem, height is measured from the bottom of a ball to the ground. A ball is released from a height of two feet. After each bounce, the ball rises to a height that is one quarter of its previous height. What total vertical distance does the ball travel by the time it strikes the ground for the fourth time?

44. When a team scores during a new sport, the team's score either increases by 4 or by 9. If team X beats team Y, the total points scored during a game by both teams is less than 25, and team Y's score is nonzero, how many different outcomes are possible for the final scores of a game?

45. The sum of the digits of a two-digit number is 11. When the digits of the number are reversed, the new number is smaller than the original number by 27. What is the number?

46. An author is deciding whether to set the list price of a digital book at $1 or $3. If the list price is $1, the royalty will be 25 cents, but if the list price is $3, the royalty will be $2. From experience, the author estimates that the book will sell 4000 copies if it is priced at $3. If the book is instead priced at $1, how many copies will need to sell for the lower price to be more profitable?

47. A customer purchases seven apples and four bananas. There is no sales tax. The total is \$6.75. Each banana costs \$0.90. How much does each apple cost?

48. Finding 80% of 6 is equivalent to finding 64% of which number?

49. Students begin an exam when a clock displays 8:00. The exam is supposed to last 45 minutes. Unfortunately, the clock is running fast. The clock adds one minute to the displayed time every 54 seconds (which is 6 seconds earlier than usual). How much time do the students lose as a result of the clock's performance?

50. Five times the perimeter of a square is 140 units. What is the area of the square?

51. Three pennies are tossed into the air. If they are all fair coins, what is the likelihood that exactly two of the pennies will land heads up?

52. The value of a stock was initially \$90. First the value of the stock increased by twenty percent. Next the value of the stock decreased by twenty percent. What is the final value of the stock? Explain why the answer is **<u>not</u>** \$90.

53. Three consecutive even numbers have a product of 7920. What are the three numbers?

54. A store is having a sale on balloons where a customer who buys three will receive two free. How many balloons does a customer need to pay for in order to have a total of 40 balloons?

55. Reggie was born today. Yolanda is currently 36 years old. In how many years will Yolanda be five times as old as Reggie?

56. How many different 4-digit numbers can be formed using only 3's, 5's, and 8's? Examples of valid numbers include 3583 and 5555.

57. Eileen has a quiz tomorrow morning at 8:15 a.m. It is currently 9:48 p.m. How much time does Eileen have before her quiz begins? Express your answer in hours and minutes.

58. Dan normally earns $16 per hour. However, if Dan works more than 40 hours in one week, Dan earns 50% more for each hour over 40 hours. How many hours does Dan need to work in one week in order to earn $1000 for the week (prior to taxes and deductions)?

59. A device rotates 15 times per second. How many times does it rotate in 2.5 minutes?

60. The average value of 13.8, 15.3, and a third number is 14.7. What is the third number?

61. An editor found 8 typos on the first page that she read. Each time the editor read a new page, she found 3 more typos than she had found on the previous page. The editor read a total of 12 pages. What was the total number of typos that she found?

62. A rectangle is seven and five-sixths inches long and is six and three-fourths inches wide. Find the area and the perimeter of the rectangle.

63. On a day when 1 euro equates to 1.15 US dollars, how many euros would 920 US dollars be worth?

64. How many different 6-character license plates (for a single state) can have 3 uppercase letters followed by a three-digit number if the letter I and the digit zero are not used? Examples of valid license plates are ODD135 and XYZ383.

65. A contractor hires 32 people to work the first day. Each day after the first, half as many people work compared to the day before. This pattern continues through the last day, when just a single person works. Each person earns $50 for one day of work. How much money must the contractor pay in total for the people hired to work?

66. The sum of the ages of five people is currently 82. What will the sum of their ages be 15 years from now?

67. Four customers paid $8 each, seven customers paid $12 each, ten customers paid $16 each, six customers paid $20 each, and three customers paid $24 each. What amount did a customer pay on average?

68. If you subtract 48 from a number and then multiply the difference by eight-thirds, the number that you get is the same as the original number. What is the number?

69. A small triangle has an area of 17 square inches. A large triangle has 6 times the base and 7 times the height of the small triangle. What is the area of the large triangle?

70. A girl read a book that is 175 pages long in 7 hours. How much time does it take her to read one page, on average? Express your answer in minutes and seconds.

71. One-half of Bill's age equals one-third of Will's age. Will is 7 years older than Bill. How old is each person?

72. A jug contains 2 gallons and 3 quarts of lemonade. All of the lemonade is to be poured out of the jug and into bottles. Each bottle can hold 1.65 pints. How many bottles are needed? Note: 1 gallon equates to 4 quarts and 1 quart equates to 2 (US liquid) pints.

73. The first 20 DVD's cost \$3.60 each. The 21st thru 50th DVD's each cost \$0.75 less than the first 20 DVD's. The 51st thru 100th DVD's each cost \$0.75 less than the 21st thru 50th DVD's. Any additional DVD's beyond the 100th each cost \$0.75 less than the 51st thru 100th DVD's. How much does it cost to purchase 128 DVD's? There is no sales tax.

74. A music service offers rates of \$3 per day, \$15 per week, or \$595 per year. If a customer uses the service for one week, how much will the customer save with the weekly rate compared to the daily rate? If a customer uses the service for one year, how much will the customer save with the annual rate compared to the weekly rate? Note: There are approximately 52 weeks in one year.

75. Base nine uses the digits 0-8 only. In base nine, the number 10 follows the number 8. The number 10 in base nine is the same as the number 9 in base ten. The number 20 in base nine follows the number 18. The number 20 in base nine is the same as the number 18 in base ten. What does 12×43 equal in base nine? Also, how would the same problem and answer be written in base ten?

76. A boy has three unusual dice. The unusual feature is that the sides are all even numbers: 0, 2, 4, 6, 8, and 10. If the three dice are fair, what is the most likely sum when all three are rolled? Also, what is the likelihood that this sum will occur?

77. A family drives a car along an interstate highway beginning at mile marker 178. The mile markers increase up to 331 at the state line. After crossing the state line, the mile markers start over at 1 and increase. At mile marker 189, the car exits the highway. How far did the car travel along the highway?

78. How many degrees correspond to three and one-half revolutions? Note: One revolution corresponds to 360°.

79. Adding a number to 11 gives the same sum as adding the negative of the number to 45. What is the number?

80. An exam has 192 questions. A student took 32 minutes to complete the first 24 questions. At this rate, how much time would the student need to complete the exam? Express your answer in hours and minutes.

81. Find the sum of all of the positive whole numbers greater than one and less than 72 which evenly divide into 72.

82. A puppy is 13 weeks and 4 days old. A kitten is one-fifth as old as the puppy. How old is the kitten? Express your answer in weeks and days.

83. A total of forty students are arranged in three groups. Two of the groups have the same number of students. The third group has half as many students as each of the other two groups. How many students are in each group?

84. Ten raised to what power equals a billion?

85. Four squares are arranged in a row. Each square is red, blue, or green. Two adjacent squares may not be the same color. The first and third squares must be the same color. How many different ways are there to arrange the four squares?

86. A cat ate one-half of its food. An hour later, the cat ate one-third of the food that remained. An hour after that, the cat ate one-fourth of the food that was still in its bowl. This pattern continued until the cat ate one-eighth of the food that was still in its bowl. At this point, what percent of the food did the cat eat all together?

87. There are seven girls and five boys in a class. After the class, every girl shakes hands once with each of the other girls, and every boy shakes hands once with each of the other boys. However, no boy shakes hands with a girl. What is the total number of handshakes?

88. There are seven girls and five boys in a class. After the class, every girl shakes hands once with each of the boys. No boy shakes hands with another boy, and no girl shakes hands with another girl. What is the total number of handshakes?

89. The length and width of a rectangle are measured. The results are recorded as 6.4 ± 0.2 m and 4.5 ± 0.3 m, where the plus or minus ($\pm$) notation indicates the worst-case error in the measurements. For example, 6.4 ± 0.2 could be any value from $6.4 - 0.2$ to $6.4 + 0.2$. Find the minimum and maximum possible values for the perimeter.

90. The length and width of a rectangle are measured. The results are recorded as 6.4 ± 0.2 m and 4.5 ± 0.3 m, where the plus or minus ($\pm$) notation indicates the worst-case error in the measurements. For example, 6.4 ± 0.2 could be any value from $6.4 - 0.2$ to $6.4 + 0.2$. Find the minimum and maximum possible values for the area.

91. A man travels 1 mile north, 1 mile west, 2 miles south, 2 miles east, 3 miles north, 3 miles west, 4 miles south, 4 miles east, 5 miles north, 5 miles west, 6 miles south, 6 miles east, 7 miles north, 7 miles west, 8 miles south, 8 miles east, 9 miles north, 9 miles west, and 5 miles south. How far and which way should the man travel in order to return to his starting point in a straight line?

92. A company advertises its product online with a pay-per-click service. The advertisement generates 800 clicks. Each click costs the company 75 cents. If the company makes a profit of $3 per product sold, what percent of sales does the company need from the clicks in order for the advertisement to yield a short-term profit? Note: Compare the number of sales needed to the number of clicks to determine the percent.

93. The SI units of speed are meters per second. The SI units of acceleration are meters per second squared (m/s^2). If you square the SI units of speed and divide this by the SI units of acceleration, what do you get?

94. A store sells 27 batteries for $8. How many batteries can you buy for $56? There is no sales tax.

95. Four hundred dollars are put into a savings account that earns interest at a rate of 3%, compounded once per year. There are no other transactions. There are no taxes or fees. What is the account balance after six years?

96. Find the smallest whole number that is an even multiple of both 36 and 48.

97. A kid is playing a pattern recognition game. When the boy types 1, a screen displays 7. When the boy then types 2, 3, 4, 5, 6, 7, 8, 9, 10, and 11, the screen displays the numbers tabulated below. Predict which number will be displayed on the screen if the boy types 131.

What the boy types	1	2	3	4	5	6	7	8	9	10	11
What is displayed	7	32	11	2	7	32	11	2	7	32	11

98. Two fractions have a sum of five thirds and a difference of one sixth. What are the fractions?

99. The letter a is worth 1 point, b is worth 2 points, c is worth 3 points, etc., such that z is worth 26 points. Divide by the number of letters in a word to find its score. For example, for the word "sun," the s is worth 19 points, the u is worth 21 points, and the n is worth 14 points. Add these points and divide by the number of letters in the word (3) to find that "sun" has a score of 18. What is the longest word that you can think of which has a score higher than 20? Note: There are many acceptable answers to this question (but remember to divide by the number of letters in the word).

100. A circular pizza is cut into five equal slices. Find the angle of each slice. Note: You want the angle involving the vertex that lies at the center of the circle.

101. One bag contains brown balls and pink balls. Thirty-six percent of the balls in that bag are brown. Another bag contains twice as many balls as the first bag. All of the balls in the second bag are pink. If all of the balls from the two bags are mixed together, what percent of the balls will be brown?

102. Find the greatest common factor of 180 and 255.

103. Nishi can type 75 words per minute. Lee can type 50 words per minute. If Nishi and Lee both type on separate computers, about how much time would it take for them to type a total of one thousand words?

104. Multiplying a negative number by negative three gives the same answer as subtracting the negative number from twenty-two. What is the number?

105. Jean has 40 more monthly payments to make on her car loan. Jean's next payment is due in November. In which month will Jean make her last payment? Jean makes all of her payments on time and pays the exact amount that is due.

106. Find the smallest whole number that is an even multiple of both 56 and 84.

107. A customer purchases three-fourths of a pound of rice and two-thirds of a pound of beans. There is no sales tax. The total is $1.70. The rice costs $1.60 per pound. What is the price per pound of the beans?

108. A total of 84 blocks are arranged in three stacks. The second stack has half as many blocks as the first stack, and the third stack has half as many blocks as the second stack. How many blocks are in each stack?

109. Seven discs numbered 1 thru 7 are placed in a bag. Two of the discs are removed at random. What is the likelihood that the numbers on the two discs add up to 9?

110. Seven discs numbered 1 thru 7 are placed in a bag. One disc is removed at random and then it is placed back in the bag. One more disc is removed at random. What is the likelihood that the numbers on the two discs add up to 9? Explain why the answer is different from the answer to the previous problem.

111. A parcel of land currently has a value of $30,000. If its value increases by 20% every decade for three decades, what will its value be at the end of the third decade?

112. The area of a circle divided by its circumference is equal to 5.4 feet. What is the diameter of the circle?

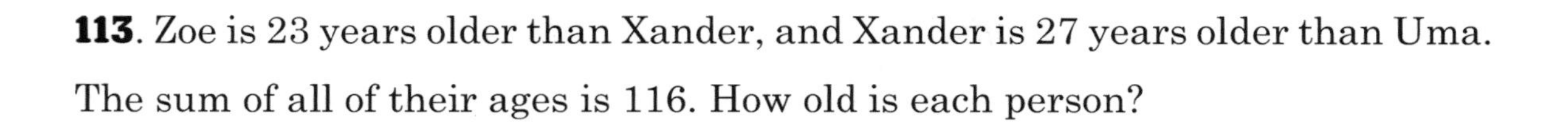

113. Zoe is 23 years older than Xander, and Xander is 27 years older than Uma. The sum of all of their ages is 116. How old is each person?

114. Subtracting a number from zero gives the same answer as adding one-half of the number to negative sixty. What is the number?

115. A model is scaled such that a distance of 3 centimeters in the model represents an actual distance of 25 meters. In the model, a building is 7.5 centimeters tall. What is the actual height of the building?

116. The tens digit of a two-digit number is twice the units digit. When the sum of the digits is subtracted from the two-digit number, the result is 72. What is the two-digit number?

117. At the same moment, a yellow dot and purple dot appear on a computer monitor and begin moving toward one another with constant speed. The yellow dot travels 7 pixels per second while the purple dot travels 5 pixels per second. The two dots are initially 720 pixels apart. When the two dots meet, how much farther has the yellow dot traveled compared to the purple dot?

118. Find the sum of all of the multiples of 9 between 10 and 100.

119. Initially, Emily had twice as many origami sheets as Drew and Drew had three times as many origami sheets as Connie. After Emily gave 28 origami sheets to Connie and Emily gave 4 origami sheets to Drew, all three students had the same number of origami sheets. What total number of origami sheets did these three students have?

120. What number cubed equals one millionth?

121. Over the course of one complete week, beginning at midnight leading into early Sunday morning and ending at midnight just as Saturday ends, a woman has been awake for a total of 106 hours and 45 minutes. For how much time has she slept each day, on average? Express your answer in hours and minutes.

122. A farmer plants 124 seeds in 8 rows. Each of the first 7 rows have the same number of seeds, but the eighth row has three-fourths as many seeds as each of the other rows. How many seeds are in each row?

123. Base three uses the digits 0, 1, and 2 only. In base three, the numbers begin 1, 2, 10, 11, 12, 20, 21, 22, 100, 101, 102, etc. What does 111×222 equal in base three? Also, how would the same problem and answer be written in base ten?

124. A rectangle initially has a perimeter of 80 units. The rectangle is stretched so that its length doubles while its width remains the same. The perimeter of the stretched rectangle is 128 units. What are the length and width of the original rectangle?

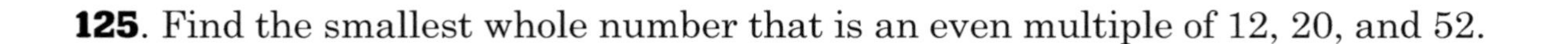

125. Find the smallest whole number that is an even multiple of 12, 20, and 52.

126. The sum of the digits of a three-digit number is 12. The second digit is two more than the third digit. When the three-digit number is reversed, it is larger than the original number by 198. What is the number?

127. A quiz has 4 multiple choice questions. Each question has 4 choices, but only one of the choices is correct for each question. If the answers are selected at random without reading the questions, what is the likelihood of scoring 75% or higher?

128. The SI unit of distance is the meter. The SI units of acceleration are meters per second squared (m/s^2). If you divide the SI unit of distance by the SI units of acceleration and then take the square root of that, what do you get? Express your answer in its simplest form.

129. Brand A's batteries last an average of 60 hours each. Brand B's batteries last an average of 45 hours each. Brand A sells an 8-pack of batteries for $4. Brand B sells their batteries in packs of 12. How cheap does brand B's 12-pack need to be in order for it to be more economical than brand A's 8-pack?

130. A cube has a surface area of 384 square units. What is the volume of the cube?

131. Two decimals have a sum of 1.2 and a difference of 0.28. What are the decimals?

132. A long staircase has 1000 steps numbered 1 thru 1000. Fernando is at the bottom, just below step 1. Gilbert is at the top on step 1000. At the same time, Fernando begins climbing the stairs with constant speed and Gilbert begins descending the stairs with a different constant speed. Fernando climbs 8 stairs in the time it takes for Gilbert to descend 17 stairs. Which stair will Fernando and Gilbert be on when they meet?

133. A number plus the same number squared plus the same number cubed is equal to 1,010,100. What is the number?

134. Marco has $135. Marco wishes to buy a printer that costs $200. Marco has a mystery coupon that will reveal a percent discount when he attempts to make the purchase. The tax rate is 8%. What minimum percent discount does Marco need in order to be able to purchase the printer?

135. Find the greatest common factor of 108 and 450.

136. An engineer designs an "unfair" coin that comes up heads two out of three times, on average. If three such unfair coins are tossed at the same time, what is the likelihood that all three coins will land heads up? If three fair coins are tossed at the same time, what is the likelihood that all three coins will land heads up?

137. Imagine five separated dots on a piece of paper. What is the maximum number of line segments that can be drawn such that every line segment connects a different pair of dots and such that no two line segments cross? Position the dots as needed to get the maximum number of line segments satisfying these conditions, provided that no two dots touch and that any given pair of points is not joined by more than one line segment.

138. A can of paint can cover 48 square yards of surface area. A woman needs to paint 2500 square feet of surface area. How many cans of paint should she buy? How many additional square feet could she paint with the paint that is left over? Notes: Assume that only a single coat of paint is needed. Note that 48 square yards has different units than 2500 square feet.

139. If you multiply a number by three, square this product, and then add two dozen, you get 600. What is the number?

140. A bucket contains three times as many washers as bolts. A second bucket contains half as many washers as bolts. The sum of the number of washers and the number of bolts is the same for each bucket. All of the washers and bolts are poured into a single barrel (which was previously empty). Find the ratio of washers to bolts in the barrel.

141. How many hundred-dollar bills are one million nickels worth?

142. Initially, a train is traveling and there are 243 passengers on board. Every time the train stops at a train station, one-third of the passengers exit the train and nobody boards the train. When passengers finish exiting at the fifth train station, how many passengers are still aboard the train?

143. Find the average value of a half, a third, and a fourth.

144. Katherine and Larry are driving go-karts around a circular track that has a circumference of 240 meters. Katherine drives with a constant speed of 24 m/s and Larry drives with a constant speed of 18 m/s. They start side by side. How much time elapses when Katherine passes Larry by exactly one lap?

145. Veronica is now twice as old as Isabel. Also, Veronica is now three times as old as Isabel was four years ago. How old is each person now?

146. An auditorium has a total of 1700 seats arranged in two sections. The right section has fifty rows and the left section has thirty rows. The rows in the right section each have the same number of seats. The rows in the left section also each have the same number of seats, but a row in the left section has 40% fewer seats than a row in the right section. How many seats are in each row of each section?

147. Find the sum of the prime factors of 1190.

148. A bag contains 5 brown blocks, 4 yellow blocks, and 3 green blocks. All of the blocks are identical except for their colors. If two of the blocks are selected at random, what is the likelihood that both blocks will be green?

149. What is 36% of five-eighths of 2.4?

150. Angie and Steve only have quarters and dimes. Angie has three times as many quarters as Steve, and Angie has half as many dimes as Steve. Together, they have fifty percent more dimes than quarters. The total value of all of their money is $38.40. How much money does each person have?

151. A store has small, medium, and large gloves in the ratio 4:9:5. There are a total of 144 gloves. How many gloves of each size are there?

152. How many centimeters are there in one kilometer?

153. A family baked a cake. The daughter ate one-eighth of the cake. The son ate one-sixth of what was left. The mother ate one-fifth of what was left. The father ate one-fourth of what was left. How much cake remained after everyone ate?

154. In an experiment, when the force is 2 N, the distance is 720 cm. When the force is 5 N, the distance is 288 cm. When the force is 9 N, the distance is 160 cm. Predict what the distance will be when the force is 36 N.

155. A truck travels in a straight line with a constant speed of 18 m/s. The truck is 12 meters long. A bug flies in the same direction as the truck, but parallel to the truck's path (so that the bug does not hit the truck), with a constant speed of 26 m/s. From the instant the bug is level with the back of the truck until the instant that the bug passes the front of the truck, how much time passes?

156. If the bug in the previous question traveled in the opposite direction of the truck (instead of the same direction as the truck), how much time would pass from the instant the bug is level with the front of the truck until the instant that the bug passes the back of the truck?

157. A blueprint is scaled such that 0.25 inches corresponds to 3 feet. A room in the blueprint has a width of 0.75 inches and a length of 1.25 inches. What is the area of the actual room in square feet?

158. There are 50 regions on a map. Initially, 34% of the regions are red, 42% of the regions are blue, and 24% of the regions are yellow. If 4 blue regions are changed to red and 2 yellow regions are also changed to red, what percent of the regions will be each color?

159. A woman gave birth to twins. Each twin gave birth to triplets. Each triplet gave birth to quadruplets. A family get-together is attended by the woman, all of the twins, all of the triplets, and all of the quadruplets (but nobody else). How many people are at the family get-together?

160. How many different 6-digit numbers have one 4, one 5, three 7's, and one 9? An example of a valid number is 977,457.

161. The SI unit of distance is the meter. The SI units of acceleration are meters per second squared (m/s^2). If you multiply the SI unit of distance by the SI units of acceleration and then take the square root of that, what do you get? Express your answer in its simplest form.

162. A sheet of paper is initially 8.5 inches wide and 11 inches long. The paper is cut in half, making two rectangles with one side that measures 8.5 inches. One half is discarded in a trash can. The paper is similarly cut three more times so that each half of the sheet has one side that matches the shorter dimension of the previous rectangle. Each time, one half is discarded in a trash can. What fraction of the original area does the final piece of paper have?

163. Twenty minus the square root of a number equals five plus the square root of the same number. What is the number?

164. Gina and Hank are initially next to one another on a circular jogging track. At the same moment, Gina begins traveling with a constant speed of 5 m/s in the clockwise direction while Hank begins traveling with a constant speed of 3 m/s in the counterclockwise direction. The circumference of the track is one kilometer. When they first meet up, how far has each person traveled?

165. Five bottles of water and three bowls of fruit cost $19.50. Four bottles of water and seven bowls of fruit cost $31.70. There is no sales tax. How much does it cost for one bottle of water? How much does it cost for one bowl of fruit? Assume that all of the bottles of water cost the same. Assume that all of the bowls of fruit cost the same.

166. A rectangle is three times as long as it is wide. The area of the rectangle is 108 square centimeters. What is the perimeter of the rectangle?

167. Jordan and Kim have a jigsaw puzzle. If Jordan worked on the jigsaw puzzle alone, Jordan would finish in two hours. If Kim worked on the jigsaw puzzle alone, Kim would finish in one hour and twenty minutes. If Jordan and Kim work on the jigsaw puzzle together, about how much time will it take for them to finish?

168. Find the greatest common factor of 210, 336, and 504.

169. A cord is initially eight and one-third feet long. The cord is cut such that two and one-half feet are removed from one end and discarded in the trash. The remaining section of cord is then cut into five equal pieces. How long is each piece? Express your answer in feet and inches.

170. Money is put into a savings account that earns interest at a rate of 4%, compounded once per year. There are no other transactions. There are no taxes or fees. After two years, the interest earned is $51. Find the initial account balance and the final account balance.

171. If the lengths of all of the edges of a square are increased by forty percent, by what percent is the area of the square increased?

172. A small brick has half the weight of a medium brick. A medium brick has half the weight of a large brick. One small brick, one medium brick, and one large brick are placed on a scale. The combined weight is 28 pounds. How much does each brick weigh?

173. Beginning at 8:00 a.m. and finishing at 7:59 p.m. on the same day, how many of the times displayed on a standard digital clock have at least one digit that is a 2?

174. Short, medium, and long strands of lights contain a total of 586 lightbulbs. A medium strand contains four more lightbulbs than a short strand. A long strand contains six more lightbulbs than a medium strand. There are 8 short strands, 10 medium strands, and 15 long strands. How many lightbulbs are on each type of strand?

175. When Haley first rode her bike along a track with constant speed, her time was 75 seconds. When Haley next rode her bike along the same track with a different constant speed, her time was 60 seconds. Haley rode 8 m/s faster on her second attempt than she had on her first attempt. Determine Haley's speed for each attempt.

176. Eight times a number equals eighteen divided by the same number. What is the number?

177. How many cubes with an edge length of 2 inches need to be glued together in order to form the shape of a rectangular box measuring 6 in. × 8 in. × 10 in.? The final shape must be solid (without any cavities inside or on the surface).

178. Initially, a store had twice as many laptops as desktops. Three-fifths of the laptops and two-thirds of the desktops were sold. Of the laptops and desktops that were not sold, what is the ratio of laptops to desktops?

179. On the same day, a woman invested \$2000 in stocks and \$5000 in a savings account. The savings account paid interest at a rate of 3% per year. After one year, she earned a combined amount of \$300 from her two investments. What interest rate did she earn from the stocks?

180. If you add one to a number and square the sum, what you get is 25 more than if you square the number without first adding one. What is the number?

181. A pond has three separate drains. If only the left drain is unplugged, the pond drains in 40 minutes. If only the middle drain is unplugged, the pond drains in 48 minutes. If only the right drain is unplugged, the pond drains in 60 minutes. If all three drains are unplugged, about how much time does it take for the pond to drain?

182. At its highest speed, a particular scooter can travel 72 kilometers in one hour. How many meters can the scooter travel in one second?

183. The ratio of the edge lengths of two cubes is 3:8. Find the ratio of the volumes of the cubes. Also find the ratio of the surface areas of the cubes.

184. Yvonne is taking four math tests, each with a maximum possible score of 100. Yvonne earned scores of 96 and 92 on her first two tests. What is the lowest score that she can earn on one of the next two tests and still be able to have an average score of at least 90?

185. Dividing 126 by a number gives the same quotient as dividing the number by 14. What is the number?

186. A boy tied 66 ropes to a group of trees. Each end of a rope is tied to one tree. Every tree in the group is connected to every other tree by a single rope. There is exactly one rope connecting any pair of trees in the group. How many trees are in the group?

187. Given three random people, what is the likelihood that at least two were born in the same month? Note: The question does not care whether or not they were born in the same year. Only the month matters.

188. A right-circular cylinder has a circumference of 2.5 feet. The cylinder rolls without slipping with a constant speed such that its center of mass travels 1.5 feet per second for a total time of 3 minutes. How many complete rotations does the cylinder make during this time?

189. Currently, one divided by Olivia's age plus one divided by Patrick's age is equal to one divided by Quince's age. Olivia is 28 years old and Quince is 12 years old. How old is Patrick?

190. Five hundred dollars are put into a savings account that earns interest at a fixed rate, compounded once per year. There are no other transactions. There are no taxes or fees. After two years, the account balance is $551.25. What is the interest rate?

191. Sally and Tom have a total of 180 identical small squares. Sally and Tom each happen to have exactly the number of small squares needed to arrange their own small squares in the shape of a larger (solid) square. Sally's large square has twice as many small squares along each edge as Tom's large square. How many small squares does each person have?

192. A clock is correctly set to 12:00. When the clock displays 12:24, it is really 12:30 because the clock is running slow. When the clock displays 12:48, it is really 1:00. When the clock first displays 4:00, what times is it really?

193. In an experiment, when the width is 2 meters, the height is 7 meters. When the width is 3 meters, the height is 26 meters. When the width is 4 meters, the height is 63 meters. When the width is 5 meters, the height is 124 meters. Predict what the height will be when the width is 8 meters.

194. A square has an edge length of 24 centimeters. The square is split into two rectangles. One of the rectangles has three times the area of the other rectangle. Find the ratio of the perimeters of the rectangles.

195. Average speed is defined as the total distance traveled divided by the elapsed time. A robot travels with a constant speed of 30 m/s along a straight line, and then travels with a constant speed of 20 m/s in the opposite direction until it reaches its starting point, immediately stopping at its starting point. Explain why the average speed must be **less than** 25 m/s. Also, determine the value of the average speed.

196. Base twelve uses the digits 0-9 and the characters A and B. In base twelve, after 9 come A, B, and then 10. The numbers A, B, and 10 in base twelve are the same as ten, eleven, and twelve, respectively, in base ten. What does 2A × B2 equal in base twelve? Also, how would the same problem and answer be written in base ten?

197. In one bag of rocks, twenty-five percent of the rocks are shiny. In a second bag of rocks that has twice as many rocks as the first bag, forty percent of the rocks are shiny. When all of the rocks from both bags are poured into a single box (which was previously empty), what percent of all of the rocks are shiny?

198. A round table has six equally spaced chairs. Four boys and two girls sit down at the table according to a seating chart. If the seating arrangement is random, what is the likelihood that both girls will be sitting next to one another?

199. A boat has 140 gallons of water inside of it. A father uses a bucket to remove 50 gallons of water from the boat every hour. His son uses another bucket to remove 75 gallons of water from the boat every hour. A leak in the bottom of the boat allows 20 gallons of water into the boat every hour. If the father and son work continually, how much time will it take for the boat to be momentarily empty of water?

200. In a cartoon, a squirrel steals an acorn from one dog and begins to run towards another dog. At the same moment, each dog begins traveling towards the other dog with a constant speed of 4 m/s. The two dogs are initially 200 m apart. Each time the squirrel reaches a dog, the squirrel immediately reverses direction. The squirrel always travels with a constant speed of 15 m/s. What is the total distance that the squirrel travels before the squirrel can no longer run away from a dog by reversing its direction?

IDEA CENTER

The Art of Solving Word Problems

To solve a word problem, the first steps are to identify the information that is given in the problem and to identify what the problem is asking you to find. It may help to circle or highlight key words or information in the problem, and it may help to make a list of quantities that you know.

Look for signal words that may help you connect the words to the math. For example, the words "sum," "total," "difference," and "product" usually signal a specific arithmetic operation. However, you really need to learn to comprehend the language rather than rely on signal words. Why? The English language is complex, allowing numerous ways to effectively convey the same information. For example, instead of using "difference" to indicate subtraction, a problem may use language like "left over," "lost," "remained," "taken away," or several alternatives.

An important way to improve your ability to understand what a word problem is asking for and how the wording relates to the math is to practice solving a variety of word problems. This book provides two hundred different problems for you to gain experience through practice. After you solve or attempt to solve a problem, check the answer key in the back of the book. If you made a mistake, spend some time thinking about what you did wrong and how you can avoid that mistake in the future. Learning from your mistakes can help make you a better problem-solver.

Problem-Solving Techniques

There is not just one way to correctly solve a word problem, and different word problems tend to be solved in different ways. However, there are some general ideas that are helpful for solving a variety of word problems.

- If the problem can be visualized, draw a picture. It is amazing how often a labeled diagram helps students realize something that they had not noticed before drawing the picture. It is like turning on a light to help see better in the dark.
- Circle, highlight, or list the information that you are given or that you have already figured out. Some or all of this information is needed in the solution. You want to be able to find it easily.
- Try to figure out which kind of math relates to the problem. Are there any signal words like “ratio” or “remainder”? Does the problem relate to a particular topic, like fractions or like geometry?
- Are there any formulas that may apply to the problem? A formula may help to relate information involving speed, interest rates, average values, perimeter, area, or volume, for example.
- Be organized. Write information down so that you can easily find it. Make your process and ideas clear.
- Be brave. Don’t be afraid to try an idea out and see where it leads. You never know until you try.
- If you know algebra, you should use it. Although the word problems in this book can be solved without algebra, for some of the problems, algebra offers an advantage. Define what each variable represents. Write down one or more equations involving the unknowns. Apply algebra to solve for an unknown.

Formula for Average (Arithmetic Mean)

To find the average value of N numbers, add up the numbers and divide by N.

$$\text{average} = \text{sum of numbers} \div N$$

Example: Find the average of 18, 25, and 35.

Solution: $(18 + 25 + 35) \div 3 = 78 \div 3 = 26$.

Geometry Formulas

To find the perimeter of a rectangle, multiply the length by two, multiply the width by two, and add these terms together.

$$P = 2 \times L + 2 \times W$$

To find the area of a rectangle, multiply the length by the width.

$$A = L \times W$$

To find the area of a triangle, multiply the base by the height and divide by two.

$$A = b \times h \div 2$$

To find the volume of a cube, raise the edge length to the power of three.

$$V = L^3$$

To find the volume of a rectangular box, multiply the dimensions together.

$$V = L \times W \times H$$

To relate the diameter of a circle to its radius, multiply the radius by two.

$$D = 2 \times R$$

To find the circumference of a circle, multiply two, π, and the radius, where π is approximately equal to 3.14.

$$C = 2 \times \pi \times R$$

To find the area of a circle, multiply π times the radius squared.

$$A = \pi \times R^2$$

Example: What is the area of a triangle that has a base of 6 and height of 8?

Solution: The base is b = 6 units. The height is h = 8 units. The area of a triangle is one-half of the base times the height.

$$A = b \times h \div 2 = 6 \times 8 \div 2 = 24 \text{ square units}$$

Example: What is the perimeter of a rectangle that measures 9 × 7?

Solution: The length is L = 9 units and the width is W = 7 units.

$$P = 2 \times L + 2 \times W = 2 \times 9 + 2 \times 7 = 18 + 14 = 32 \text{ units}$$

Rate Equations

When an object travels with constant speed, the rate equation can be expressed in three equivalent forms:

$$\text{speed} = \text{distance} \div \text{time}$$

$$\text{distance} = \text{speed} \times \text{time}$$

$$\text{time} = \text{distance} \div \text{speed}$$

The units can help. Distance is measured in units of length, such as meters or feet. Time is measured in units like hours or seconds. Speed has a ratio of units like m/s or mph (miles per hour).

Example: A car travels with a constant speed of 40 mph for 30 minutes. How far does the car travel?

Solution: The speed is 40 mph. The time is 30 minutes. However, the units of the time are not consistent with the units of the speed. Observe that mph has an hour in it (mph = miles per hour), whereas the time is in minutes. Convert 30 minutes to 0.5 hours, knowing that 1 hour = 60 minutes. Now that the time and speed have compatible units, we may use one of the rate equations. The middle rate equation allows us to solve for the distance.

$$\text{distance} = \text{speed} \times \text{time} = 40 \text{ mph} \times 0.5 \text{ hr} = 20 \text{ miles}$$

Two (or More) Objects or People

If there are multiple objects or people in a problem, it may help to make a table to keep track of the information for each object or person.

Example: Two monkeys are separated by 200 meters on a long ladder. The top monkey begins descending with a constant speed of 7 m/s at the same time that the bottom monkey begins climbing with a constant speed of 13 m/s. When will the two monkeys meet? How far does each monkey travel during this time?
Solution: The monkeys have different speeds and travel different distances, but the time is the same for each. (The monkeys begin moving at the same moment.) The rate equation states that distance equals the product of speed and time.

	speed	time	distance
top monkey	7 m/s	t	$7 \times t$
bottom monkey	13 m/s	t	$13 \times t$

One way to solve this problem is to note that each second the two monkeys get $7 + 13 = 20$ meters closer to one another. It will thus take $200 \div 20 = 10$ seconds for the monkeys to meet.
Another way to solve this problem is to apply the rate equation and use algebra. The two distances add up to 200 meters.

$$7\,t + 13\,t = 200$$

$$20\,t = 200$$

$$t = 200 \div 20 = 10 \text{ seconds}$$

Now that we know the time, we may plug it into the rate equation to determine how far each monkey travels. The top monkey travels $7 \times 10 = 70$ meters. The bottom monkey travels $13 \times 10 = 130$ meters. One way to help check the answers is to note that 70 m + 130 m = 200 m.

Example: John is currently 12 years older than Hanna. Six years ago, John was twice as old as Hanna. How old is each person now?

Solution: One way to solve this problem is to note that the difference between their ages is 12 years now and six years ago the difference was also 12 years. Go back six years, when John was twice as old as Hanna. If John was twice as old as Hanna and John was also 12 years older than Hanna, six years ago Hanna must have been 12 years old and John must have been 24 years old. To find their current ages, add 6 years: John is currently $24 + 6 = 30$ years old and Hanna is currently $12 + 6 = 18$ years old.

	now	6 years ago
John	H + 12	H + 6
Hanna	H	H – 6

Another way to solve this problem is to apply algebra. Let H represent Hanna's age now. John's age now is then $H + 12$. Six years ago, Hanna's age was $H - 6$ and John's age was $H + 12 - 6 = H + 6$. Six years ago, John was twice as old as Hanna. Express this with an equation:

$$H + 6 = 2(H - 6)$$

Apply the distributive property: $a(b - c) = ab - ac$.

$$H + 6 = 2H - 12$$

Subtract H from both sides. Add 12 to both sides. Combine like terms.

$$12 + 6 = 2H - H$$

$$18 = H$$

Currently, Hanna is 18 years old and John is $H + 12 = 18 + 12 = 30$ years old.

Check the answers:

- John is 12 years older than Hanna: $30 - 12 = 18$.
- 6 years ago, John was $30 - 6 = 24$ years old and Hanna was $18 - 6 = 12$.
- 6 years ago, John was twice as old as Hanna: $24 = 2 \times 12$.

Compound Interest

Interest equals the product of the principal (which is the beginning balance) and the interest rate (in decimal form): $I = P \times r$. The balance after one time period equals the principal plus the interest: $A = P + I$.

If the interest is left in a savings account, the interest gets compounded with each additional time period. For example, for an interest rate that is computed once per year, $I_1 = P_1 \times r$ and $P_2 = P_1 + I_1$ after the first year and $I_2 = P_2 \times r$ and $A = P_2 + I_2$ at end of the second year.

For multiple time periods, it may help to express the interest rate in a different form. Note, for example, that if a principal of \$80 increases by 4%, it is simpler to multiply \$80 × 1.04 to determine the new balance than it is to first compute \$80 × 0.04 and then add this to \$80. Either way, you get the same new balance. Note that \$80 × 1.04 = \$83.20 and \$80 × 0.04 + \$80 = \$3.20 + \$80 = \$83.20. If the principal is unknown and the interest is compounded over multiple time periods, working with $1 + r$ instead of just r may make the solution simpler.

For many time periods, there is a compound interest formula:

$$A = P \times (1 + r/n)^{nt}$$

Here, t is the number of time periods (such as 5 years) and n is the compounding frequency (which is the number of times that interest is applies per time period). For an annual interest rate that is applied once per year, t would be the number of years and $n = 1$, such that $A = P \times (1 + r)^t$. Note that t is an exponent.

Example: \$800 is put in a savings account that earns 2.5% interest per year. What is the balance after one year?

Solution: The principal is P = \$800. The interest rate is r = 2.5% ÷ 100% = 0.025. The interest is $I = P \times r$ = \$800 × 0.025 = \$20. The final balance is $A = P + I$ = \$800 + \$20 = \$820.

Example: \$100 is put in a savings account that earns 6% interest per year. What is the balance after two years?

Solution: The original principal is P_1 = \$100. The interest rate is r = 6% ÷ 100% = 0.06. The interest after one year is $I_1 = P_1 \times r$ = \$100 × 0.06 = \$6. The balance after one year is $P_2 = P_1 + I_1$ = \$100 + \$6 = \$106. The interest earned for the second year is $I_2 = P_2 \times r$ = \$106 × 0.06 = \$6.36. The final balance is $A = P_2 + I_2$ = \$106 + \$6.36 = \$112.36.

Another way to solve this problem is:

$$P_2 = P_1 \times 1.06 = \$100 \times 1.06 = \$106$$

$$A = P_2 \times 1.06 = \$106 \times 1.06 = \$112.36$$

Yet another solution is:

$$A = P \times (1 + r)^t = \$100 \times (1 + 0.06)^2 = \$100 \times (1.06)^2 = \$100 \times 1.1236 = \$112.36$$

Working Together

If two or more people work together to complete a task, determine what fraction of the task each person completes per time period. Add these fractions together to determine how much work is done together each time period. Then you can determine how much time it will take to complete the work, like the example that follows.

Example: Liz could eat a bag of popcorn in 6 minutes by herself. Meg could eat a bag of popcorn in 12 minutes by herself. If they share a bag of popcorn, about how much time will it take for Liz and Meg to finish the popcorn?

Solution: Liz eats 1/6 of the bag each minute. Meg eats 1/12 of the bag each minute. Together, they eat 1/6 + 1/12 = 2/12 + 1/12 = 3/12 = 1/4 of the bag each minute. It will take about 4 minutes for Liz and Meg to eat the bag of popcorn. It should make sense that the answer is less than 6 minutes.

Ratios and Proportions

A ratio expresses a fixed relationship, separating two numbers with a colon like 5:8. A ratio is basically a fraction. For example, the ratio 5:8 can be expressed as the fraction 5/8.

Note that a ratio may be expressed part-to-part, part-to-whole, or whole-to-part. For example, suppose that there are 11 girls and 9 boys in a class. The ratio of girls to boys, 11:9, is part-to-part, whereas the ratio of girls to students, 11:20, is part-to-whole.

A proportion sets two ratios equal to one another. For example, if the ratio of paperbacks to hardcovers on a shelf is 5:2 and there are 75 paperbacks on the shelf, the number of hardcovers can be determined by making a proportion. The proportion requires the fraction 5/2 to equal 75 divided by the number of hardcovers. Which number can you divide 75 by to make 5/2? Since $75 \div 5 = 15$, we can multiply $2 \times 15 = 30$ to determine that $5/2 = 75/30$. There are 30 hardcovers on the shelf. (Multiply 5 and 2 each by 15 to see that 5/2 equals 75/30.)

Example: The ratio of red balloons to yellows balloons is 3:4. The balloons only come in these two colors. There are 140 balloons. How many red balloons and how many yellow balloons are there?

Solution: Note that the ratio is part-to-part, whereas 140 involves the whole. If there are 3 red balloons for every 4 yellow balloons, there are 3 red balloons for every 7 balloons of any color. The ratio 3:7 is part-to-whole. Multiply the part-to-whole ratio, 3/7, by the whole, 140, to determine the number of red balloons: $140 \times 3/7 = 60$. Subtract the number of red balloons from the whole to determine that the number of yellow balloons is $140 - 60 = 80$.

Check the answer: The ratio of red balloons to yellow balloons is 60:80, which reduces to 3:4 if you divide 60 and 80 each by 20.

Counting and Likelihood

The letters of the word FUN can be arranged 6 different ways:

FNU, FUN, NFU, NUF, UFN, UNF

The letters of the word MATH can be arranged 24 different ways:

AHMT, AHTM, AMHT, AMTH, ATHM, ATMH

HAMT, HATM, HMAT, HMTA, HTAM, HTMA

MAHT, MATH, MHAT, MHTA, MTAH, MTHA

TAHM, TAMH, THAM, THMA, TMAH, TMHA

The different ways of ordering the letters of a word are called permutations. If every letter is different, the number of permutations equals a factorial. We use the notation N! to represent a factorial. This means to multiply successively smaller whole numbers until reaching one. For example, $3! = 3 \times 2 \times 1 = 6$ and $4! = 4 \times 3 \times 2 \times 1 = 24$.

If a word has repeated letters, divide by a factorial for each letter that repeats. For example, the letters of the word PUPPY can be arranged $5! \div 3! = 120 \div 6 = 20$ different ways. We divided by 3! because the word PUPPY has 3 P's.

PPPUY, PPPYU, PPUPY, PPUYP, PPYPU,

PPYUP, PUPPY, PUPYP, PUYPP, PYPPU,

PYPUP, PYUPP,UPPPY, UPPYP, UPYPP,

UYPPP, YPPPU, YPPUP, YPUPP, YUPPP

As another example, the digits of the five-digit number 12121 can be arranged $5! \div (3! \times 2!) = 120 \div (6 \times 2) = 120 \div 12 = 10$ different ways. We divided by both 3! and 2! because there are 3 ones and 2 twos.

11122, 11212, 11221, 12112, 12121, 12211, 21112, 21121, 21211, 22111

Example: How many different ways can the letters of PROOF be arranged?

Solution: Divide by 2! because there are 2 O's: $5! \div 2! = 120 \div 2 = 60$.

To determine the likelihood (or probability) of an event occurring, first make a list of the possible outcomes, like the example below. If the list is very long, make a list of sample outcomes and use the sample to help determine the likelihood.

Example: Five tiles are numbered 1 thru 5. The five numbered tiles are placed in a bag. If two tiles are drawn from the bag at random, what is the likelihood that the tiles will have a sum of 7?

Solution: There are 20 different ways to draw 2 tiles from the bag.

1 and 2, 1 and 3, 1 and 4, 1 and 5, 2 and 1, 2 and 3, 2 and 4,
2 and 5, 3 and 1, 3 and 2, 3 and 4, 3 and 5, 4 and 1, 4 and 2,
4 and 3, 4 and 5, 5 and 1, 5 and 2, 5 and 3, 5 and 4

There are 4 ways for the sum to be 7:

$$2 + 5 = 7,\ 3 + 4 = 7,\ 4 + 3 = 7,\ 5 + 2 = 7$$

There are 4 ways for the sum to be 7 out of 20 different ways. The likelihood is 1 out of 5 that the sum will be 7 (since the ratio 4:20 reduces to 1:5).

Change of Base

We normally express numbers in base ten, but it is possible to work with other bases besides base ten. For example, base six uses the digits 0-5 only. In base six, the number 10 follows the number 5. Then number 10 in base six is the same as the number 6 in base ten. The first 18 numbers in base 6 are shown below.

Base 10	1	2	3	4	5	6	7	8	9	10	11	12
Base 6	1	2	3	4	5	10	11	12	13	14	15	20

Base 10	13	14	15	16	17	18	19	20	21	22	23	24
Base 6	21	22	23	24	25	30	31	32	33	34	35	40

Base 10	25	26	27	28	29	30	31	32	33	34	35	36
Base 6	41	42	43	44	45	50	51	52	53	54	55	100

When converting between base six and base ten, the powers of six are special. The numbers $6^1 = 6$, $6^2 = 6 \times 6 = 36$, $6^3 = 6 \times 6 \times 6 = 216$, and so on in base ten correspond to the numbers 10, 100, 1000, and so on in base six. This makes it easy to convert a number from base ten to another base. See the example below.

Base 10	$6^1 = 6$	$6^2 = 36$	$6^3 = 216$	$6^4 = 1296$
Base 6	10	100	1000	10,000

Example: What does 41×54 equal in base six? Also, how would this problem and answer be written in base ten?

Solution: From the previous table, 41 in base six corresponds to 25 in base ten and 54 in base six corresponds to 34 in base ten. The problem 41×54 in base six corresponds to 25×34 in base ten. In base ten, it is easy to determine that $25 \times 34 = 850$. Now we just need to convert 850 from base ten to base six. The way to do this is to rewrite 850 in terms of powers of six, which are $6^1 = 6$, $6^2 = 6 \times 6 = 36$, $6^3 = 6 \times 6 \times 6 = 216$, etc. We can take out as many as three 216's from 850, which leaves $850 - 3 \times 216 = 850 - 648 = 202$. We can take out as many as five 36's from 202, which leaves $202 - 5 \times 36 = 202 - 180 = 22$. We can take out as many as three 6's from 22, which leaves $22 - 3 \times 6 = 22 - 18 = 4$. Putting all of this together, we get the following in base ten:

$$850 = 3 \times 216 + 5 \times 36 + 3 \times 6 + 4$$

The table above shows that 216 in base ten is equivalent to 1000 in base six, 36 in base ten is equivalent to 100 in base six, and 6 in base 10 is equivalent to 10 in base six. Plug these numbers in the line above to rewrite 850 in base six.

$$3 \times 1000 + 5 \times 100 + 3 \times 10 + 4 = 3000 + 500 + 30 + 4 = 3534$$

The number 3534 in base six is equivalent to 850 in base ten. The multiplication problem $41 \times 54 = 3534$ in base six is equivalent to the multiplication problem $25 \times 34 = 850$ in base ten.

A Few Math Notes

The greatest common factor (GCF) of two whole numbers is the largest whole number that evenly divides into each. For example, the GCF of 32 and 40 is 8 because $32 = 8 \times 4$ and $40 = 8 \times 5$.

The least common multiple (LCM) of two whole numbers is the smallest whole number that is evenly divisible by each. For example, the LCM of 6 and 8 is 24 because $6 \times 4 = 24$ and $8 \times 3 = 24$.

The prime factorization of a whole number refers to the prime numbers that can be multiplied together to make the whole number. For example, the prime factorization of 200 is $5^2 \times 2^3$ since $5^2 = 5 \times 5$, $2^3 = 2 \times 2 \times 2 = 8$, and $25 \times 8 = 200$.

Adding a negative number to a positive number is equivalent to subtraction. For example, $12 + (-5) = 12 - 5 = 7$. Subtracting a number from a negative number makes a more negative number. For example, $-8 - 4 = -12$.

Multiplying a positive number by a negative number makes a negative product. For example, $7 \times (-3) = -21$. When multiplying two negative numbers, the minus signs cancel out. For example, $-7 \times (-3) = 21$.

An exponent (or power) indicates repeated multiplication. For example, 5^3 has three fives multiplied together: $5^3 = 5 \times 5 \times 5 = 125$.

The square root of a number has two possible answers since it may be positive or negative. For example, the square root of 9 can be 3 or -3 because $3^2 = 3 \times 3 = 9$ and $(-3)^2 = (-3) \times (-3) = 9$ also. The negative answer may be important.

To add or subtract fractions, make a common denominator. For example, to add $3/4 + 1/6$, the lowest common denominator (LCD) is twelve:

$$3/4 + 1/6 = 9/12 + 2/12 = 11/12$$

To reduce a fraction, divide the numerator and denominator each by the GCF. For example, 8/12 reduces to 2/3 if you divide 8 and 12 each by 4.

To find the reciprocal of a fraction, interchange the roles of the numerator and denominator. For example, the reciprocal of 2/3 is 3/2. To find the reciprocal of a whole number, divide one by the number. For example, the reciprocal of 5 is 1/5. If a fraction has 1 in the numerator, the reciprocal is a whole number. For example, the reciprocal of 1/5 is 5.

To divide by a fraction, multiply by its reciprocal. For example, 4/3 ÷ 5/9 is the same as 4/3 × 9/5 = 36/15 = 12/5. (In the last step, we divided 36 and 15 each by 3 in order to reduce 36/15 to 12/5.)

Note that a fraction is equivalent to division. For example, 42/6 equals 7 since 42 ÷ 6 = 7. Similarly, 7/4 is equivalent to 1.75 because 7 ÷ 4 = 1.75.

To convert a decimal to a percent, multiply by 100%. For example, 0.325 equates to 32.5%.

To convert a percent to a decimal, divide by 100%. For example, 63% equates to 0.63.

There are two ways to deal with a percent increase or decrease. For example, if $60 increases by 25%, you can first determine that 0.25 × $60 = $15 and then add $15 to $60 to get $75. However, it is more efficient to multiply 1.25 × $60 = $75. Similarly, if $80 decreases by 30%, you can first determine that 0.3 × $80 = $24 and then subtract to get $80 – $24 = $56. However, it is more efficient to multiply 0.7 × $80 = $56. (In the first case, we added 1 + 0.25 to get 1.25, and in the second case we subtracted 1 – 0.3 to get 0.7.)

In the context of fractions, decimals, and percents, the word "of" is often (but not always) used to mean multiplication. For example, 20% of 9 is 0.2 × 9 = 1.8.

ANSWER KEY

1. Final answer: 5:8 is the ratio of vowels to consonants for the sentence.

Notes: The ratio 20:32 reduces to 5:8. Divide 20 and 32 each by 4 to get 5:8.

There are $4 + 1 + 3 + 1 + 2 + 1 + 3 + 1 + 1 + 3 = 20$ vowels.

There are $5 + 2 + 2 + 1 + 4 + 1 + 7 + 2 + 3 + 5 = 32$ consonants.

Determine = 5 consonants + 4 vowels

the = 2 consonants + 1 vowel

ratio = 2 consonants + 3 vowels

of = 1 consonant + 1 vowel

vowels = 4 consonants + 2 vowels

to = 1 consonant + 1 vowel

consonants = 7 consonants + 3 vowels

for = 2 consonants + 1 vowel

this = 3 consonants + 1 vowel

sentence = 5 consonants + 3 vowels

2. Final answer: 2160 is how many aardvarks are equivalent to 3 elephants.

Note: 3 elephants = 3×15 giraffes = $3 \times 15 \times 8$ zebras = $3 \times 15 \times 8 \times 6$ aardvarks

3. Final answer: 21 pages are read in total.

Notes: Although 75 minus 74 equals one, if you read pages 74-75, you actually read two pages, not one. Similarly, if you read pages 17-22, you read pages 17, 18, 19, 20, 21, and 22, which is 6 pages, even though 22 minus 17 equals 5. You have to add one each time after you subtract. The answer is

$$(22 - 17 + 1) + (51 - 39 + 1) + (75 - 74 + 1) = 6 + 13 + 2 = 21$$

A sure way to see that there are 21 pages is to list them. See the next page.

17, 18, 19, 20, 21, 22, 39, 40, 41, 42, 43, 44, 45, 46, 47, 48, 49, 50, 51, 74, 75

For curious minds: You can see a related concept by looking at a human hand. If your hand has 5 fingers, it has 4 spaces between the fingers. If you subtract 22 minus 17 to get 5, you are getting the 5 spaces between the pages. You need to add one to get all 6 pages read. Look at it another way. You start on page 17. To reach page 22, you advance 5 pages. This is what you get when you subtract. To also account for the starting page, you add one. To find the number of pages that you advanced, just subtract. To find the number of pages that you read, subtract and then add one to account for the first page. This counting issue often comes up in various ways in math competitions.

4. Final answers: 181 and 97 have an average of 139 and a difference of 84.

Check the answers: The average value is (97 + 181) ÷ 2 = 278 ÷ 2 = 139.

The difference is 181 – 97 = 84.

5. Final answers: Fred is 27 years old and Bonnie is 9 years old.

Check the answers: Fred is 3 times as old as Bonnie: 27 = 3 × 9.

The sum of their ages is 9 + 27 = 36.

6. Final answer: 11:53 a.m. was the time when Quentin began driving.

Check the answer: It takes 7 minutes to go from 11:53 a.m. to noon.

Since 75 – 7 = 68, this leaves 68 minutes left: 60 minutes take you from noon to 1:00 p.m. and the last 8 minutes take you from 1:00 p.m. to 1:08 p.m.

7. Final answer: 9 inches is the height of the orange rectangle.

Notes: If you double the length of the green rectangle and also triple the height of the green rectangle, the new rectangle (which is orange) will have six times the area because 2 × 3 = 6. When we triple the height of the green rectangle, we get 3 × 3 inches = 9 inches.

Try it with numbers: Suppose that the length of the green rectangle is 5 inches. Then the area of the green rectangle is 5 × 3 = 15 square inches, the length of the orange rectangle is $5 \times 2 = 10$ inches, the height of the orange rectangle is $3 \times 3 = 9$ inches, and the area of the orange rectangle is $10 \times 9 = 90$ square inches. Note that 90 square inches is 6 times larger than 15 square inches. Although making up a number for the length of the green rectangle helps to check the answer with numbers, it is not a desirable way to solve the problem. A teacher would likely deduct points for solving a problem by making up numbers. One way to solve this problem without making up any numbers is to reason out the answer like we discussed in the Notes following the answer. Another way is to use algebra. For example, write $A = LH$ for the green rectangle, which becomes $A = 3L$ (since $H = 3$), such that $6A = (2L)(y)$ for the orange rectangle (since it has 6 times the area, 2 times the length, an unknown height of y). Plug $A = 3L$ into $6A = 2Ly$ to get $18L = 2Ly$, which simplifies to $9 = y$.

8. Final answer: 30 students are **<u>not</u>** wearing jackets.

Notes: 45 students are wearing jackets and 30 students are not.

There are 15 fives in 75 since $15 \times 5 = 75$.

$15 \times 3 = 45$ equals the number wearing jackets, and $15 \times 2 = 30$ equals the number who are not wearing jackets. Alternatively, $75 - 45 = 30$.

There are 3 students wearing jackets for every 2 students not wearing jackets (such that 3 out of every 5 students are wearing jackets; note that $3 + 2 = 5$). Since 45 are wearing jackets, this means that $(2/3) \times 45 = 30$ are not.

9. Final answer: 96 is the number of times the butterfly flaps its wings in 12 s.

Notes: One minute equals 60 seconds. The butterfly flaps its wings 480 times in 60 seconds. Divide $480 \div 60 = 8$ to determine that the butterfly flaps its wings at a rate of 8 times per second. Multiply 8 times per second by 12 seconds to find that the butterfly flaps its wings 96 times in 12 seconds.

This problem involves a proportion: 480 is to 60 as 96 is to 12. The two ratios (480:60 and 96:12) both equal 8:1.

10. Final answer: 2 m/s to the south is the boy's speed in the last trip.

Notes: The rate equation has three equivalent forms:

speed = distance ÷ time

distance = speed × time

time = distance ÷ speed

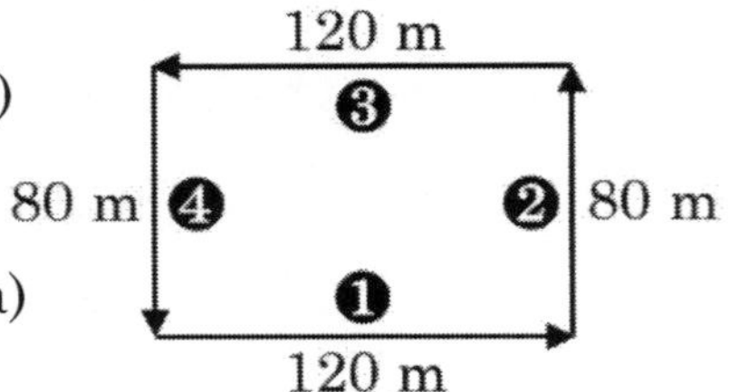

Trip 1 (east): 5 m/s × 24 s = 120 m (middle formula)

Trip 2 (north): 80 m ÷ 16 s = 5 m/s (top formula)

Trip 3 (west): 120 m ÷ 3 m/s = 40 s (bottom formula)

Trip 4 (south): 80 m ÷ 40 s = 2 m/s (top formula)

The boy traveled along a rectangle. This is how we know that the boy traveled south in Trip 4.

Trips 1-3 took 24 + 16 + 40 = 80 s. The entire trip took 2 minutes, which equates to 120 s. Subtract 80 s from 120 s to determine that Trip 4 took 40 s.

11. Final answer: $13.66 is how much each girl receives; the remainder is 2 cents.

Check the answer: $13.66 × 3 = $40.98 and $40.98 + $0.02 = $41.00

Note: One way to solve this problem is to do long division with 41.00 ÷ 3.

12. Final answer: $15,300 is the value of the car now.

Check the answer: $18,000 is the original value of the car.

15% of $18,000 is 0.15 × $18,000 = $2700.

Subtract $2700 from the original value to find the current value.

$18,000 – $15,300 = $2700

Note: One way to solve this problem is to divide $2700 by 0.15 and then subtract $2700. Since 0.15 equals 15/100 and since the way to divide by a fraction is to multiply by its reciprocal, $2700 divided by 0.15 is the same as

$2700 × 100 ÷ 15 = $18,000

Now subtract $2700 from the original value of $18,000:

$18,000 – $2700 = $15,300

13. Final answers: 13 pencils and 4 erasers, with 8 cents left over.

Check the answers: 13 pencils cost 13 × \$0.72 = \$9.36.

4 erasers cost 4 × \$0.14 = \$0.56.

$$\$9.36 + \$0.56 + \$0.08 = \$10.00$$

14. Final answer: 28 is the total number of handshakes.

Notes: The handshake problem is a classic type of question that appears in many math competitions in different forms. A common mistake is to multiply 8 by 7, but to forget to divide by 2 to correct for double counting. Visually, this problem relates to a picture called a complete graph, where every vertex connects to every other vertex, as shown below.

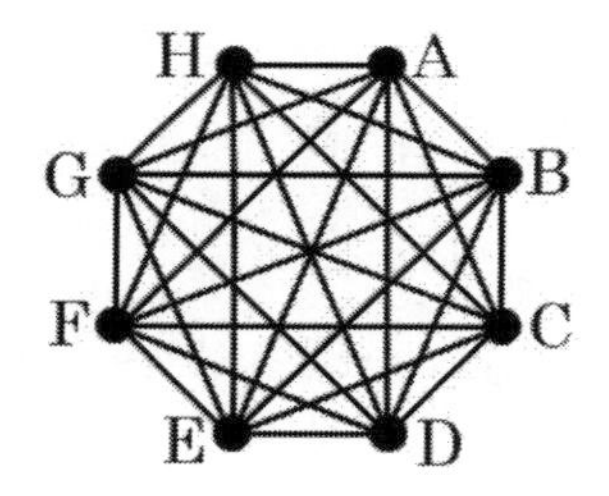

Solution: Method 1: Multiply 8 by 7 and divide by 2 to correct for double counting:

$$8 \times 7 \div 2 = 56 \div 2 = 28$$

Method 2: Add 7 + 6 + 5 + 4 + 3 + 2 + 1 = 28.

Notes: Name the people A, B, C, D, E, F, G, and H. Note that A shakes hands with 7 other people: B, C, D, E, F, G, and H. Similarly, B shakes hands with 7 other people: A, C, D, E, F, G, and H. However, A shaking hands with B is the same handshake as B shaking hands with A. Note that C also shakes hands with 7 other people: A, B, D, E, F, G, and H. In this case, we already counted A shaking hands with C and B shaking hands with C. You should see that A shakes hands with 7 other people, B shakes hands with 6 people in addition to A, C shakes hands with 5 people in addition to A and B, etc. The solution in Method 2 above uses this line of thinking.

For the solution in Method 1, we begin by multiplying 8 by 7 to get 56 because everybody shakes hands with every other person, but then we divide by 2 since every handshake was counted twice. When we get 56, the problem is that A shaking hands with B and B shaking hands with A were both counted, and all other handshakes were counted twice (such as A with C and C with A, B with C and C with B, etc.). We divide 56 by 2 to correct for double counting.

15. Final answer: 24 is how many times the pendulum will oscillate back and forth in one minute.

Check the answer: The pendulum oscillates 24 times. Each oscillation takes 2.5 seconds. The total time is 24×2.5 seconds = 60 seconds.

16. Final answers: 48 and 19 have a sum of 67 and a difference of 29.

Check the answers: $48 + 19 = 67$ and $48 - 19 = 29$.

17. Final answers: Maria is 15 years old and Victor is 45 years old.

Check the answers: Victor is 3 times as old as Maria since $45 = 15 \times 3$.

Ten years ago, Victor was $45 - 10 = 35$ and Maria was $15 - 10 = 5$. Ten years ago, Victor was 7 times as old as Maria since $35 = 5 \times 7$.

18. Final answers: Iris has 45 quarters and Oscar has 135 dimes.

Check the answers: The value of Iris's money is $45 \times \$0.25 = \$11.25$.

The value of Oscar's money is $135 \times \$0.10 = \$13.50$.

Oscar has \$2.25 more than Iris because $\$13.50 - \$11.25 = \$2.25$.

19. Final answer: 20 is how many 2×2 tiles are needed.

Check the answers: Method 1: The area of the rectangle is $8 \times 10 = 80$ square units. The area of one tile is $2 \times 2 = 4$ square units. Divide 80 by 4 to get 20.

Method 2: The 8×10 rectangle is 4 tiles wide and 5 tiles high. Multiply 4 by 5 to get 20.

20. Final answer: 60 different 5-character codes can be made using two J's, one Q, one 4, and one 8.

Notes: One way to find the answer is to first multiply 5 times 4 times 3 times 2 times 1, and then divide by 2 to correct for double counting of the J's.

Method 1: The Q can be in 5 different positions. Once the Q is set, the 4 can be in 4 different positions. Once the Q and 4 are set, the 8 can be in 3 different positions. Once the Q, 4, and 8 are set, the two J's automatically fill the two remaining positions. This makes $5 \times 4 \times 3 = 60$ different character codes.

Method 2: A 5-letter word with 5 different letters, like "smile," can have its letters arranged $5 \times 4 \times 3 \times 2 \times 1 = 120$ different ways. Why? If there are two different letters, like "is," we can make $2 \times 1 = 2$ different arrangements (is and si). If there are three different letters, like "sim," we can make $3 \times 2 \times 1 = 6$ different arrangements (sim, smi, mis, msi, ims, ism). If there are four different letters, like "slim," we can make $4 \times 3 \times 2 \times 1 = 24$ different arrangements. To see this, note that each of 4 letters can come first, and for whichever letter comes first, there are 6 ways to arrange the 3 letters that follow it. For example, when s comes first, we get slim, slmi, smil, smli, silm, and siml. If there are five different letters, like "smile," this generalizes to $5 \times 4 \times 3 \times 2 \times 1 = 120$. However, if there are repeated letters, we need to divide to correct for double counting. In this case, 4JQ8J has two J's. If the word "smile" had been "smele," there would only have been 60 different ways to arrange the letters instead of 120. When the two vowels are different, it makes a difference which vowel is an i and which is an e. For example, smile is different from smeli. However, if both vowels are the same, swapping the vowels makes no difference. For example, smele is the same as smele (swapping the e's makes the same word).

Method 3: The permutation formula $N! \div M!$ is a simple way to apply Method 2. Note that N! means N times (N – 1) times (N – 2), etc. until you reach one. For example, $3! = 3 \times 2 \times 1 = 6$. For this problem, $N = 5$ (since there are 5 characters) and $M = 2$ (since J appears 2 times). See the following page.

$$5! \div 2! = (5 \times 4 \times 3 \times 2 \times 1) \div (2 \times 1) = 120 \div 2 = 60$$

Notes: If two different letters had repeated, we would have a factorial (that is what M! is called) for each repeated letter. For example, the word "banana" has 6 letters (3 a's, 2 n's, and 1 b), so we would get $6! \div (3! \times 2!)$.

21. Final answer: 8.5 percent is the rate of the sales tax.

Check the answer: Divide by 100% to convert 8.5% to a decimal: 8.5% = 0.085.

Multiply \$16 by 0.085 to find the amount of the sales tax: \$16 × 0.085 = \$1.36.

Add \$1.36 and \$16 to find the total: \$16 + \$1.36 = \$17.36.

Alternatively, add 0.085 to 1 to get 1.085 and multiply \$16 by 1.085.

One way to solve the problem is to divide \$17.36 by \$16 and subtract one.

Another way to solve the problem is to subtract \$16 from \$17.36 to get \$1.36, and then divide that by \$16.

22. Final answers: Initially, Natalie had 78 binder clips and Claire had 42. After Natalie gave one-third of her binder clips to Claire, Natalie had 52 binder clips and Claire had 68.

Check the answers: Initially Natalie had 78 – 42 = 36 more binder clips than Claire. In the end Natalie had 68 – 52 = 16 fewer binder clips than Claire. Also note that 78 – 78/3 = 78 – 26 = 52 and 42 + 78/3 = 42 + 26 = 68.

23. Final answer: 9 rooms are numbered from 140 thru 172 counting by 4's.

Note: This problem involves the counting issue that we discussed at length in the solution to Problem 3. It may help to review that solution.

Check the answer: For this problem, it is easy to list and count the rooms:

140, 144, 148, 152, 156, 160, 164, 168, 172

Solution: $[(172 - 140) \div 4] + 1 = (32 \div 4) + 1 = 8 + 1 = 9$

24. Final answer: 630 miles is how far the car can travel on 14 gallons of gas.

Notes: Multiply the miles per gallon by the number of gallons:

45 miles/gallon × 14 gallons = 630 miles

25. Final answer: 1360 K will be the temperature when the pressure is 17 Pa.

Notes: Multiply the pressure by 80 to find the temperature:

$$3 \text{ Pa} \times 80 = 240 \text{ K}$$

$$5 \text{ Pa} \times 80 = 400 \text{ K}$$

$$8 \text{ Pa} \times 80 = 640 \text{ K}$$

$$17 \text{ Pa} \times 80 = 1360 \text{ K}$$

It is not necessary to know any science to answer this question. You just need to be able to identify the pattern.

26. Final answers: 163, 165, and 167 are three consecutive odd numbers that add up to 495.

Notes: Five less than five hundred is $500 - 5 = 495$.

Divide 495 by 3 to find the average value: $495 \div 3 = 165$. The average value is the middle number. Subtract 2 and add 2 to find the least and greatest values. Note the key terms "consecutive" and "odd."

27. Final answer: 5 out of 36 is the probability that the sum will equal 10.

Notes: There are 5 ways for the sum to equal 10:

$$1 + 9,\ 3 + 7,\ 5 + 5,\ 7 + 3,\ 9 + 1$$

There are 36 possible outcomes for the sum of the two dice:

$$1 + 1,\ 1 + 3,\ 1 + 5,\ 1 + 7,\ 1 + 9,\ 1 + 11$$

$$3 + 1,\ 3 + 3,\ 3 + 5,\ 3 + 7,\ 3 + 9,\ 3 + 11$$

$$5 + 1,\ 5 + 3,\ 5 + 5,\ 5 + 7,\ 5 + 9,\ 5 + 11$$

$$7 + 1,\ 7 + 3,\ 7 + 5,\ 7 + 7,\ 7 + 9,\ 7 + 11$$

$$9 + 1,\ 9 + 3,\ 9 + 5,\ 9 + 7,\ 9 + 9,\ 9 + 11$$

$$11 + 1,\ 11 + 3,\ 11 + 5,\ 11 + 7,\ 11 + 9,\ 11 + 11$$

28. Final answer: 22 years old is Wendy's age now.

Check the answer: Wendy will be $22 + 6 = 28$ in six years. Eight years ago, Wendy was $22 - 8 = 14$. Observe that 28 is twice 14.

29. Final answer: 2.7 is 15% of 30% of 60.

Solution: Divide 15% and 30% each by 100% to convert them to decimals.

$$15\% = 0.15 \text{ and } 30\% = 0.3$$

To find 30% of 60, multiply 0.3 by 60. This is $0.3 \times 60 = 18$.

To find 15% of 18, multiply 0.15 by 18. This is $0.15 \times 18 = 2.7$.

30. Final answer: 15 calculators are in the top drawer.

Solution: $141 \div 18 = 7$ with a remainder of 15.

7 drawers each have 18 calculators. The remainder (15) are placed in the top drawer. Observe that $7 \times 18 + 15 = 126 + 15 = 141$.

31. Final answer: 24 seconds after the cat started (which is equivalent to 30 seconds after the mouse started), the robotic cat will catch the robotic mouse.

Note: Recall the following rate equation from the solution to Problem 10:

$$\text{distance} = \text{speed} \times \text{time}$$

Check the answer: The mouse, which started earlier, travels for a total of 30 s (since 24 s + 6 s = 30 s). The mouse, traveling 8 m/s, travels a distance of:

$$8 \text{ m/s} \times 30 \text{ s} = 240 \text{ m}$$

The cat travels for 24 s. It should make sense that the mouse spends more time traveling since the mouse started earlier than the cat. The cat, traveling 10 m/s, travels a distance of:

$$10 \text{ m/s} \times 24 \text{ s} = 240 \text{ m}$$

It should make sense that the cat and mouse travel the same distance because they start next to one another and finish when the cat catches the mouse.

Method 1: You can actually reason out the answer with just a little math. The mouse travels $8 \times 6 = 48$ m before the cat starts. When the cat starts, the cat travels 2 m/s faster than the mouse, so the cat will get 2 meters closer to the mouse each second. It will take $48 \div 2 = 24$ seconds for the cat to catch up to the mouse.

Method 2: Another way to solve this problem is to use algebra. One way to do this is to write

$$d = 8\ t_m$$

for the mouse, where t_m represents the time that the mouse travels, and write

$$d = 10\ t_c$$

for the cat, where t_c represents the time that the cat travels. If you do this, you must be careful as, for whatever reason, many students get the times mixed up in the following equation. The way to get it right is to consider the following explanation. Since the mouse starts sooner than the cat, the mouse spends more time traveling than the cat. This means that t_m must be larger than t_c. Since t_m is larger than t_c, we must add the 6 seconds to t_c. (The smaller time plus 6 is equal to the larger time.)

$$t_m = t_c + 6$$

Now you can replace t_m with $t_c + 6$ in the equation $d = 8\ t_m$. Then set the right-hand sides of the two distance equations equal to solve for t_c.

32. Final answer: $24.50 is the cost to purchase 56 markers.

Notes: Since 56 ÷ 8 = 7, it will cost 7 times as much money to buy 56 markers as it costs to buy 8 markers. Multiply 7 × $3.50.

This problem involves a proportion: 56 is to 8 as $24.50 is to $3.50. The two ratios (56:8 and $24.50:$3.50) both equal 7:1.

33. Final answer: 130 miles is the actual distance between the two cities.

Check the answer: 0.5 inches corresponds to 20 miles, so 1 inch corresponds to 40 miles and 3 inches corresponds to 120 miles. Divide 0.5 and 20 each by two to determine that 0.25 inches corresponds to 10 miles. Since 3.25 = 3 + 0.25, 3.25 inches corresponds to 120 + 10 = 130 miles.

A simpler solution is to start with 3.25 ÷ 0.5 = 6.5. Then multiply 6.5 by 20 to get 6.5 × 20 = 130.

34. Final answer: 3 to 2 is the ratio of odd numbers to even numbers.

Check the answer: There are 36 odd numbers and 24 even numbers in the list. The ratio 36:24 reduces to 3:2 if you divide 36 and 24 each by 12. Note that 36 odd numbers and 24 even numbers satisfies the information given: The list has 36 + 24 = 60 numbers, and there are 36 – 24 = 12 more odd numbers than even numbers.

Notes: The problem states that the list contains 60 numbers, but it does not say which numbers these are. The numbers can **not** be 1 thru 60, obviously, because if that were the case, there would not be 12 more odd numbers than even numbers.

35. Final answer: 75.0 centimeters (or simply 75 centimeters) is the length of the metal rod at room temperature.

Notes: The metal rod is 2% longer than it would be at room temperature due to thermal expansion. Divide 2% by 100% to convert it to a decimal: 2% = 0.02. Divide 76.5 cm by 1.02 to find the length of the metal rod at room temperature: 76.5 cm ÷ 1.02 = 75.0 cm = 75 cm. You can do this without using a calculator if you realize that 76.5 equals 153/2 and that 1.02 equals 51/50.

From a physical perspective, 2% is rather extreme, but the numbers are easier to work with than if it had been 0.1% or less.

Check the answer: 2% of 75 cm equals 0.02 × 75 cm = 1.5 cm. This is how much the metal rod has expanded. Add 1.5 cm to 75 cm to get 76.5 cm.

36. Final answer: 78.03 centimeters is the length of the wooden rod at room temperature.

Notes: In contrast to the previous problem, in this problem the meterstick is longer than normal rather than the rod being measured. This has the opposite effect; it actually makes the wooden rod seem shorter than usual. Unlike the previous problem, this time we need to multiply: 76.5 cm × 1.02 = 78.03 cm.

37. Final answer: 9.8 seconds is how much time the fourth student has left.

Check the answer: 11.2 s + 10.6 s + 10.5 s + 9.8 s = 42.1 s.

38. Final answer: December 3 is the earliest and January 7 is the latest.

Notes: October has 31 days. From Oct. 29, one day is Oct. 30, two days is Oct. 31, three days is Nov. 1, four days is Nov. 2, five days is Nov. 3, six days is Nov. 4, and one week is Nov. 5. Two weeks is Nov. 12, three weeks is Nov. 19, and four weeks is Nov. 26. November has 30 days. Count 7 more days for the fifth week: Nov. 27, Nov. 28, Nov. 29, Nov. 30, Dec. 1, Dec. 2, and Dec. 3. This makes Dec. 3 the end of the fifth week. The sixth week is Dec. 10, the seventh week is Dec. 17, the eighth week is Dec. 24, and the ninth week is Dec. 31. December has 31 days. Count 7 more days for the tenth week: Jan. 1, Jan. 2, Jan. 3, Jan. 4, Jan. 5, Jan. 6, and Jan. 7. This makes Jan. 7 the end of the tenth week. It is possible to solve this problem more efficiently. We took the time to spell it out for you to help you feel "convinced" that these are the correct answers.

39. Final answer: 60 is the total number of coins that Raj has.

Check the answer: Raj has 15 quarters, 15 dimes, 15 nickels, and 15 pennies. This makes a total of 15 + 15 + 15 + 15 = 60 coins. The total value of Raj's coins equals

$$15 \times \$0.25 + 15 \times \$0.10 + 15 \times \$0.05 + 15 \times \$0.01$$
$$= \$3.75 + \$1.50 + \$0.75 + \$0.15 = \$6.15$$

One way to solve this problem is to note that $0.25 + $0.10 + $0.05 + $0.01 = $0.41. Next divide $6.15 ÷ $0.41 = 15. Finally, multiply 15 × 4 = 60.

40. Final answer: 144 pens have blue ink

Check the answer: The black pens come in 36 groups of 5 since 180 = 36 × 5. The blue pens come in 36 groups of 4 since 144 = 36 × 4.

This problem involves a proportion: 180 is to 5 as 144 is to 4. The two ratios (180:5 and 144:4) both equal 36:1.

41. Final answer: \$3.00 (or simply \$3) was the average cost of a ticket.

Notes: It would be incorrect to add \$2.70 to \$3.60 and divide by 2. Why? To find the average, you must add all of the numbers. Since 60 children's tickets were sold and 30 adult tickets were sold, there are $60 + 30 = 90$ numbers to add. We can do this efficiently though. For example, since 60 of the tickets cost \$2.70, we can multiply \$2.70 by 60 to quickly add up the cost of the children's tickets. This is basically a weighted average, yet this problem can be solved correctly without knowing anything about weighted averages.

Solution: $(\$2.70 \times 60 + \$3.60 \times 30) \div 90 = (\$162 + \$108) \div 90 = \$270 \div 90 = \$3$.

42. Final answer: 17 years ago, Pete was three times as old as Tyrone.

Check the answers: 17 years ago, Tyrone was $29 - 17 = 12$ years old and Pete was $53 - 17 = 36$ years old. Observe that $36 = 12 \times 3$.

43. Final answer: 3.3125 feet is the total vertical distance that the ball travels before it strikes the ground for the fourth time.

Note: Since height is measured from the bottom of the ball to the ground, we do not need to know the diameter of the ball.

Solution: The ball drops 2 feet and strikes the ground for the first time.

The ball rises $2 \div 4 = 0.5$ feet.

The ball drops 0.5 feet and strikes the ground for the second time.

The ball rises $0.5 \div 4 = 0.125$ feet.

The ball drops 0.125 feet and strikes the ground for the third time.

The ball rises $0.125 \div 4 = 0.03125$ feet.

The ball drops 0.03125 feet and strikes the ground for the fourth time.

$$2 + 0.5 + 0.5 + 0.125 + 0.125 + 0.03125 + 0.03125$$
$$= 2 + 1 + 0.25 + 0.0625 = 3.3125$$

44. Final answer: 14 different outcomes are possible for which the scores satisfy the criteria specified in the problem.

Notes: A single team could score 4 points, $4 + 4 = 8$ points, 9 points, $4 + 4 + 4 = 12$ points, $9 + 4 = 13$ points, $4 + 4 + 4 + 4 = 16$ points, $9 + 4 + 4 = 17$ points, $9 + 9 = 18$ points, or $9 + 9 + 2 = 20$ points (which is equivalent to $4 + 4 + 4 + 4 + 4 = 20$), and so on. The possible scores for a single team are:

4, 8, 9, 12, 13, 16, 17, 18, 20, 21, 24, etc.

We need X's score to be higher than Y's score. We need Y's score to be nonzero. We need the scores of X and Y added together to be less than 25. The possible scores are:

8 to 4, 9 to 4, 9 to 8, 12 to 4, 12 to 8, 12 to 9, 13 to 4,
13 to 8, 13 to 9, 16 to 4, 16 to 8, 17 to 4, 18 to 4, 20 to 4

Note, for example, that 21 to 4 is not possible because 21 + 4 equals 25, which is too high. Similarly, 13 to 12 is not possible. Since we need X to beat Y, the first score always has to be higher.

45. Final answer: 74 is a two-digit number where the sum of the digits is 11 and where the reversed number is smaller than the original number by 27.
Check the answer: $74 - 27 = 47$ and $7 + 4 = 11$.

46. Final answer: At least 32,001 copies need to be sold with a list price of $1 in order for the lower list price to be more profitable.
Check the answer: If the list price is $3, the royalty per book is $2 and the total royalty is 4000 × $2 = $8000. If the list price is $1, the royalty per book is $0.25. Note that 32,000 × $0.25 = $8000. We added 1 to 32,000 to get 32,001 because the question asked for it to be "more" profitable (rather than "just as" profitable). However, the 1 on 32,001 is not really necessary since 4000 is just an estimate.

47. Final answer: $0.45 is how much each apple costs.
Check the answer: 7 × $0.45 + 4 × $0.90 = $3.15 + $3.60 = $6.75

48. Final answer: 7.5 is the number (64% of 7.5 is equivalent to 80% of 6).
Check the answer: $0.64 \times 7.5 = 4.8 = 0.8 \times 6$

49. Final answer: 270 seconds (or 4 and one-half "real" minutes) are lost.

Notes: The clock should add 1 minute to the displayed time every 60 seconds, not every 54 seconds. With this clock, every "minute" is 54 seconds long instead of 60 seconds long. The clock will display 8:45 when 54 times 45 seconds have passed: $54 \times 45 = 2430$ seconds. If the clock had been functioning normally, the exam would have lasted $60 \times 45 = 2700$ seconds. The time lost is:

$$2700 - 2430 = 270 \text{ seconds}$$

This equates to 4 and one-half "real" minutes, since $4.5 \times 60 = 270$.

50. Final answer: 49 square units is the area of the square.

Check the answer: The edge length is 7 units. The area is $7 \times 7 = 49$ square units. The perimeter is $7 + 7 + 7 + 7 = 28$ units. Observe that $5 \times 28 = 140$ units.

51. Final answer: 3 out of 8 is the probability for exactly two heads.

Solution: There are 8 possible outcomes. We will use H for heads and T for tails.

One outcome is that all three coins are heads: HHH.

There are 3 outcomes with exactly two heads: HHT, HTH, and THH. Note that the difference has to do with which coin is tails (last, middle, or first).

There are 3 outcomes with exactly one head: HTT, THT, and TTH.

One outcome is that all three coins are tails: TTT.

There are $1 + 3 + 3 + 1 = 8$ different outcomes all together.

Since 3 outcomes have exactly two heads and there are 8 different outcomes, the likelihood for exactly two heads is 3 out of 8.

52. Final answer: \$86.40 is the final value of the stock. The answer is less than \$90 because the increase uses 20% of \$90, whereas the decrease uses 20% of a larger number. The 20% does not apply to the same value each time.

Solution: Divide 20% by 100% to convert it to a decimal: 20% = 0.2.

Multiply 0.2 by \$90 to get the amount of the increase: $0.2 \times \$90 = \$18$.

Add \$18 to \$90 to get \$108.

Multiply 0.2 by \$108 to get the amount of the decrease: $0.2 \times \$108 = \$21.60$.

Subtract \$21.60 from \$108 to get \$86.40.

Notes: The first 20% applies to \$90, whereas the second 20% applies to \$108. You can write the solution more efficiently as $0.8 \times 1.2 \times \$90 = \$86.40$.

53. Final answers: 18, 20, and 22 multiply together to make 7920.

Check the answers: $18 \times 20 \times 22 = 360 \times 22 = 7920$

Notes: One way to solve this problem is to note that 7920 is approximately 8000. Recognize that $2 \times 2 \times 2 = 8$ such that $20 \times 20 \times 20 = 8000$. This shows that the three even numbers are close to 20. Since 7920 ends with 0, one of the numbers is 20. This leaves three possibilities: 16, 18, and 20 is one set, 18, 20, and 22 is another set, and 20, 22, and 24 is another set. The set 18, 20, and 22 works. If instead you use algebra to write

$$n(n + 2)(n + 4) = 7920$$

you get a rather tedious cubic equation to attempt to solve. While algebra often makes the solution to a word problem more straightforward, this problem is one of the few exceptions.

54. Final answer: A customer needs to pay for 24 balloons.

Check the answer: If a customer buys 3 balloons, the customer receives 2 free. If a customer buys $8 \times 3 = 24$ balloons, the customer receives $8 \times 2 = 16$ free. The customer will have a total of $24 + 16 = 40$ balloons.

This problem involves a proportion: 24 is to 40 as 3 is to 5. Divide 24 and 40 each by 8 to see that 24:40 reduces to 3:5. (If a customer buys 3 balloons, the customer receives 2 free, so the customer has a total of 5. Note that the ratios 24:40 and 3:5 are part-to-whole ratios. In contrast, the ratio 3:2 is part-to-part.)

55. Final answer: 9 years from today, Yolanda will be 5 times as old as Reggie.

Check the answers: In 9 years, Reggie will be 9 years old and Yolanda will be $36 + 9 = 45$. Observe that $5 \times 9 = 45$.

56. Final answer: 81 different 4-digit numbers have only 3's, 5's, and 8's.

Solution: Method 1: Each digit can be a 3, a 5, or an 8. There are three choices for each digit. Multiply 3 times 3 times 3 times 3:

$$3 \times 3 \times 3 \times 3 = 9 \times 9 = 81$$

Method 2: The first two digits can be 33, 35, 38, 53, 55, 58, 83, 85, or 88. The last two digits can also be 33, 35, 38, 53, 55, 58, 83, 85, or 88. Multiply the 9 choices for the first pair by the 9 choices for the second pair: $9 \times 9 = 81$.

Method 3: The long method is to list all 81 numbers.

3333, 3335, 3338, 3353, 3355, 3358, 3383, 3385, 3388

3533, 3535, 3538, 3553, 3555, 3558, 3583, 3585, 3588

3833, 3835, 3838, 3853, 3855, 3858, 3883, 3885, 3888

5333, 5335, 5338, 5353, 5355, 5358, 5383, 5385, 5388

5533, 5535, 5538, 5553, 5555, 5558, 5583, 5585, 5588

5833, 5835, 5838, 5853, 5855, 5858, 5883, 5885, 5888

8333, 8335, 8338, 8353, 8355, 8358, 8383, 8385, 8388

8533, 8535, 8538, 8553, 8555, 8558, 8583, 8585, 8588

8833, 8835, 8838, 8853, 8855, 8858, 8883, 8885, 8888

It is instructive to compare this solution to the solution of Problem 20. Note that the answer to Problem 20 is much smaller than $4 \times 4 \times 4 \times 4 \times 4$ because that problem restricted the number of J's, Q's, 4's, and 8's, whereas Problem 56 does not. For example, 5533 has zero 8's whereas 8888 has four 8's.

57. Final answer: 10 hours and 27 minutes.

Check the answer: It is 2 hours and 12 minutes to midnight. It is 8 hours and 15 minutes from midnight until the quiz. Add these together: $2 + 8 = 10$ hours and $12 + 15 = 27$ minutes.

58. Final answer: 55 hours is how many hours Dan needs to work.

Check the answer: $40 \times \$16 + 15 \times \$24 = \$640 + \$360 = \$1000$

59. Final answer: 2250 is how many times the device rotates in 2.5 minutes.

Notes: Multiply 2.5 minutes by 60 to convert it to seconds: $2.5 \times 60 = 150$.

Multiply 15 by 150 to find the total number of rotations: $15 \times 150 = 2250$.

60. Final answer: 15 is the third number.

Check the answer: The average is $(13.8 + 15.3 + 15) \div 3 = 44.1 \div 3 = 14.7$.

61. Final answer: 294 is the total number of typos that the editor found.

Solution: Method 1:

$$8 + 11 + 14 + 17 + 20 + 23 + 26 + 29 + 32 + 35 + 38 + 41 = 294$$

Method 2: $8 \times 12 + 3 \times 11 \times 12 \div 2 = 96 + 198 = 294$

The second method uses the formula for triangular numbers: $11 \times 12 \div 2$ is the sum of 1 thru 11. We multiplied by 3 to find the sum of 3 thru 33 in steps of 3. We added 8×12 because that is how much larger the sum of 8 thru 41 in steps of 3 is compared to the sum of 0 thru 33 in steps of 3.

62. Final answers: The area is 52 and 7/8 square inches, and the perimeter is 29 and 1/6 inches.

Solution: The area is $(7 + 5/6) \times (6 + 3/4)$. One way to do this is to convert each mixed number to an improper fraction:

$$(7 + 5/6) \times (6 + 3/4) = (47/6) \times (27/4) = 423/8 = 52 + 7/8$$

The math is somewhat simpler if you observe that 27/6 reduces to 9/2. In that case, $(47/6) \times (27/4) = (47/2) \times (9/4)$. Otherwise, you need to reduce 1269/24 by dividing 1269 and 24 each by 3.

The perimeter is

$$2 \times (7 + 5/6) + 2 \times (6 + 3/4) = 14 + 10/6 + 12 + 6/4$$

$$= 26 + 10/6 + 6/4 = 26 + 20/12 + 18/12 = 26 + 38/12$$

$$= 26 + 3 + 2/12 = 29 + 1/6$$

Note that 38/12 is equivalent to the mixed number 3 and 2/12. Also note that 2/12 reduces to 1/6. (Divide 2 and 12 each by 2 to reduce 2/12 to 1/6.)

63. Final answer: 800 euros is how much 920 US dollars would be worth.

Check the answer: Since 1 euro equates to 1.15 US dollars, multiply 800 by 1.15 to convert from euro to US dollars: $800 \times 1.15 = 920$.

Notes: The number of euros is smaller than the number of US dollars in the conversion 1 euro = 1.15 US dollars. Similarly, 800 euros = 920 US dollars has a smaller number in front of the euros. Checking the answer for consistency like this can help to prevent mistakes. (The actual conversion from euros to US dollars changes on a daily basis.)

64. Final answer: 11,390,625 different license plates can be made like that.

Solution: Since the letter I is not used, there are 25 letters to choose from (A-H and J-Z). Since the digit 0 is not used, there are 9 digits (1-9). Each letter has 25 choices and each digit has 9 choices. This is similar to the solution to Problem 56, except that we will multiply three 25's and three 9's together:

$$25 \times 25 \times 25 \times 9 \times 9 \times 9 = 11{,}390{,}635$$

Note: Different states adopt different rules for license plates, and the rules may change over time. The letter I and digit 1 are similar in appearance, as are the letter O and the digit 0. This is one reason it might be wise to avoid the letter I and the digit 0. However, even states that avoid one or both of these often do not apply the rule to personalized plates. Also, states do not allow certain words or abbreviations that might be offensive.

65. Final answer: $3150 is the total amount that must be paid for labor.

Check the answer: $(32 + 16 + 8 + 4 + 2 + 1) \times \$50 = 63 \times \$50 = \$3150$

66. Final answer: 157 years will be the sum of their ages 15 years from now.

Solution: Each person will be 15 years older, which will add $5 \times 15 = 75$ years to the total: $82 + 75 = 157$.

67. Final answer: $15.60 is the average amount earned per day.

Solution: First note that there are $4 + 7 + 10 + 6 + 3 = 30$ days in the month.

Add up the total amount earned and divide by 30 days:

$$(\$8 \times 4 + \$12 \times 7 + \$16 \times 10 + \$20 \times 6 + \$24 \times 3) \div 30$$

$$= (\$32 + \$84 + \$160 + \$120 + \$72) \div 30$$

$$= \$468 \div 30 = \$15.60 = \$15.6$$

As noted in the solution to Problem 41, this is basically a weighted average, yet this problem can be solved without knowing anything about weighted averages.

68. Final answer: 76.8 is the number.

Check the answer: Subtract 76.8 – 48 = 28.8. Multiply this difference by 8/3:

$$28.8 \times 8/3 = 230.4 \div 3 = 76.8$$

Note: If you know algebra, one way to solve this problem is:

$$(x - 48)\ 8/3 = x$$

$$8x - 384 = 3x$$

$$5x = 384$$

$$x = 384 \div 5 = 76.8$$

69. Final answer: 714 square inches is the area of the large triangle.

Solution: Method 1: If you make the base of the small triangle 6 times wider, this will increase its area to 6 × 17 = 102 square inches. If you then make the height 7 times taller, this will increase its area to 7 × 102 = 714 square inches.

Method 2. Let the area of the small triangle be $A_S = bh/2$. The area of the large triangle is $A_L = b_L h_L/2$. According to the problem, $b_L = 6b$ and $h_L = 7h$.

$$A_L = (6b)(7h)/2 = 42(bh/2) = 42A_S = 42 \times 17 = 714$$

Notes: 17 square inches gets multiplied by 6 and by 7. Although the formula for the area of a triangle involves a one-half, the one-half does not multiply the 6 and 7. Since that one-half applies to both a small triangle and a large triangle, it does not factor into the comparison. That one-half basically cancels out in the ratio of the areas: $A_L / A_S = (b_L h_L/2) \div (bh/2) = b_L h_L/bh$. If you are not convinced, try making up numbers for the base and height of the small triangle.

70. Final answer: 2 minutes and 24 seconds is the average time to read 1 page.
Solution: Multiply 7 hours by 60 to convert the time to minutes: 7 hr. = 420 min.
Divide 420 minutes by 175 pages to find the average time spent on one page:

$$420 \div 175 = 2.4$$

Divide 420 and 175 each by 35 to see that $420 \div 175 = 12 \div 5 = 2.4$. The answer is 2.4 minutes, but the problem said to state the answer as a combination of minutes and seconds. One minute equals 60 seconds, such that:

$$0.4 \text{ minutes} = 0.4 \times 60 \text{ seconds} = 24 \text{ seconds}$$

Therefore, 2.4 minutes is equal to 2 minutes and 24 seconds. (Does this seem like a long time to spend on one page? Maybe they are large pages with fine print, or maybe the subject matter is complicated, like rocket science.)

71. Final answers: Bill is 14 years old and Will is 21 years old.
Check the answers: Will is 7 years older than Bill since $14 + 7 = 21$. One-half of Bill's age is $14 \div 2 = 7$ and one-third of Will's age is $21 \div 3 = 7$.

72. Final answer: 14 bottles are needed. (At least one bottle will not be full.)
Check the answer: 2 gallons equates to 8 quarts. The jug contains 11 quarts, which equates to 22 pints. Divide $22 \div 1.65$ to get 13 and 1/3 bottles. In decimal form, you get 13.333333333... with the 3 repeating forever. Note that 13 bottles are <u>**not**</u> enough, since the answer is greater than 13.
Note: You can fill 13 bottles completely. The 14th bottle will be one-third full.

73. Final answer: \$300.30 (three hundred dollars and thirty cents) is the cost.
Solution: The first 20 DVD's cost \$3.60 each: 20 × \$3.60 = \$72.00.
The next 30 DVD's cost \$3.60 – \$0.75 = \$2.85 each: 30 × \$2.85 = \$85.50.
The next 50 DVD's cost \$2.85 – \$0.75 = \$2.10 each: 50 × \$2.10 = \$105.00.
The last 28 DVD's cost \$2.10 – \$0.75 = \$1.35 each: 28 × \$1.35 = \$37.80.
The total cost is:

$$\$72.00 + \$85.50 + \$105.00 + \$37.80 = \$300.30$$

74. Final answers: In one week, the weekly rate costs \$6 less than the daily rate. In one year, the annual rate costs \$185 less than the weekly rate.
Solution: For one week, the daily rate costs \$3 × 7 = \$21 dollars. The weekly rate saves a total of \$21 – \$15 = \$6. For one year, the weekly rate costs \$15 × 52 = \$780. For one year, the annual rate saves a total of \$780 – \$595 = \$185.

75. Final answers: 12 × 43 = 526 with all three numbers written in base 9. If all three numbers are written in base 10, it is 11 × 39 = 429.
Check the answer: 12 in base 9 is equivalent to 11 in base 10.
43 in base 9 is equivalent to 39 in base 10. Why? Base 9 "skips" 9, 19, 29, and 39, which are included in base 10. Since these 4 numbers are "skipped" in base 9, what is normally 39 in base 10 is "pushed back' to 43 in base 9.
Multiply to get 11 × 39 = 429 in base 10.
526 in base 9 is equivalent to 429 in base 10. Why? From 1-100, we "skip" 9, 19, 29, 39, 49, 59, 69, 79, 89, and 90-99, which is a total of 19 numbers. We "skip" 19 more from 101-200, 19 more from 201-300, 19 more from 301-400, and 19 more from 401-500. Then we "skip" 509 and 519. We "skip" a total of

$$5 \times 19 + 2 = 95 + 2 = 97$$

numbers. This agrees with 526 – 97 = 429.

76. Final answers: A sum equal to 14 and a sum equal to 16 are equally likely. These are the two most likely sums. (There are two answers to this question. These two sums "tie" for most likely.) The probability that the sum will be 14 is 1 out of 8, and the probability that the sum will be 16 is also 1 out of 8.
Notes: It may help to review the solution to Problem 27, which involved similar dice, but which had a somewhat simpler solution.
When the 3 dice are rolled, the sum can be any even number from 0 thru 30. There are 6 × 6 × 6 = 216 possible outcomes starting with 0 + 0 + 0 and ending with 10 + 10 + 10.

The two most likely sums are 14 and 16. There are 27 different ways for three dice to have a sum of 14:

$$0 + 4 + 10,\ 0 + 6 + 8,\ 0 + 8 + 6,\ 0 + 10 + 4$$

$$2 + 2 + 10,\ 2 + 4 + 8,\ 2 + 6 + 6,\ 2 + 8 + 4,\ 2 + 10 + 2$$

$$4 + 0 + 10,\ 4 + 2 + 8,\ 4 + 4 + 6,\ 4 + 6 + 4,\ 4 + 8 + 2,\ 4 + 10 + 0$$

$$6 + 0 + 8,\ 6 + 2 + 6,\ 6 + 4 + 4,\ 6 + 6 + 2,\ 6 + 8 + 0$$

$$8 + 0 + 6,\ 8 + 2 + 4,\ 8 + 4 + 2,\ 8 + 6 + 0$$

$$10 + 0 + 4,\ 10 + 2 + 2,\ 10 + 4 + 0$$

There are similarly 27 different ways for three dice to have a sum of 16.

The probability that the sum will be 14 equals 27 out of 216, which reduces to 1 out of 8 (if you divide 27 and 216 each by 27). Similarly, the probability that the sum will be 16 equals 1 out of 8.

77. Final answer: 342 miles is how far the car traveled along the highway.

Solution: (331 – 178) + 189 = 153 + 189 = 342

Are you wondering about the transition from mile 331 to 1? It is not an issue because there is 1 mile between mile marker 331 and mile marker 1, so there are 189 miles from mile marker 331 to mile marker 189. We do not need to do anything special regarding this transition. It is already accounted for.

Notes: This problem does not have the counting issue that we encountered in the solutions to Problems 3 and 23. Why not? In this problem, when the car travels from mile marker 178 to 179, it only travels 1 mile, whereas if we read pages 178-179, in that case we would read 2 pages. For this problem, we do not count the first mile because the car has not yet traveled any miles. With the reading problem, we count the first page because it gets read.

78. Final answer: 1260° correspond to three and one-half revolutions.

Solution: One revolution corresponds to 360°. Multiply 360° by 3.5:

$$360° \times 3.5 = 1260°$$

79. Final answer: 17 satisfies the condition stated in the problem.

Check the answer: 17 + 11 = 28 agrees with –17 + 45 = 28.

Notes: Adding negative 17 has the opposite effect of adding 17. Adding negative 17 to 45 is equivalent to subtracting 17 from 45. One way to solve this problem is to use algebra (though it is not necessary to use algebra to solve the problem):

$$n + 11 = -n + 45$$

$$2n = 34$$

$$n = 17$$

In the first step, we added n to both sides of the equation and subtracted 11.

80. Final answer: 4 hours and 16 minutes is the projected time.

Notes: Divide 192 ÷ 24 = 8. The student has only completed 1/8 of the exam so far. Multiply 32 × 8 = 256. The answer is 256 minutes, but the problem stated to express the answer as hours plus minutes. Since 4 hours = 240 minutes, this equates to 4 hours and 16 minutes.

This problem involves a proportion: 256 is to 32 as 192 is to 24. The two ratios (256:32 and 192:24) both equal 8:1.

81. Final answer: 122 is the sum of the positive whole numbers greater than 1 and less than 72 which evenly divide into 72.

Notes: 2 and 36 evenly divide into 72 because 72 ÷ 2 = 36 and 72 ÷ 36 = 2.

3 and 24 evenly divide into 72 because 72 ÷ 3 = 24 and 72 ÷ 24 = 3.

4 and 18 evenly divide into 72 because 72 ÷ 4 = 18 and 72 ÷ 18 = 4.

6 and 12 evenly divide into 72 because 72 ÷ 6 = 12 and 72 ÷ 12 = 6.

8 and 9 evenly divide into 72 because 72 ÷ 8 = 9 and 72 ÷ 9 = 8.

Since 2, 3, 4, 6, 8, 9, 12, 18, 24, and 36 evenly divide into 72, the answer is:

$$2 + 3 + 4 + 6 + 8 + 9 + 12 + 18 + 24 + 36 = 122$$

(Note that 1 and 72 are not included in the sum because the problem specified numbers that are greater than 1 and less than 72.)

82. Final answer: 2 weeks and 5 days is how old the kitten is.

Check the answer: Since there are 7 days in one week, the kitten is $2 \times 7 + 5 = 19$ days old and the puppy is $13 \times 7 + 5 = 95$ days old. Observe that $19 \times 5 = 95$.

83. Final answer: 16 students, 16 students, and 8 students are in the 3 groups.

Check the answers: Two of the groups have 16 students, one group has half as many students (since 8 is one-half of 16), and the total is $16 + 16 + 8 = 40$.

84. Final answer: 9 is the power of ten that makes one billion.

Check the answer: A power (or exponent) indicates repeated multiplication:

$$10^9 = 10 \times 10 \times 10 \times 10 \times 10 \times 10 \times 10 \times 10 \times 10 = 1{,}000{,}000{,}000$$

85. Final answer: 12 is the number of arrangements possible.

Solution: Let R = red, B = blue, and G = green. The possible arrangements are:

B G B G, B G B R, B R B G, B R B R

G B G B, G B G R, G R G B, G R G R

R B R B, R B R G, R G R B, R G R G

Notes: The first and third colors must match. The second color is different from the first and third colors. The last color is different from the first and third colors. The second and fourth colors may be the same or they may be different.

86. Final answer: 87.5% of the food was eaten in total.

Solution: Method 1: One minus 1/2 equals 1/2. One half remains.

1/3 of 1/2 equals 1/6. Subtract 1/6 from 1/2 to get 1/3. One third remains.

1/4 of 1/3 equals 1/12. Subtract 1/12 from 1/3 to get 1/4. One fourth remains.

1/5 of 1/4 equals 1/20. Subtract 1/20 from 1/4 to get 1/5. One fifth remains.

1/6 of 1/5 equals 1/30. Subtract 1/30 from 1/5 to get 1/6. One sixth remains.

1/7 of 1/6 equals 1/42. Subtract 1/42 from 1/6 to get 1/7. One seventh remains.

1/8 of 1/7 equals 1/56. Subtract 1/56 from 1/7 to get 1/8. One eighth remains.

Since 1/8 remains, this means that the cat ate a total of $1 - 1/8 = 7/8$ of the food.

Convert this to a percent to get the final answer: $7/8 = 0.875 = 87.5\%$.

Method 2: Multiply $1 \times 1/2 \times 2/3 \times 3/4 \times 4/5 \times 5/6 \times 6/7 \times 7/8 = 1/8$. Subtract 1/8 from 1 to get 7/8, which equates to 0.875 and 87.5%.

87. Final answer: 31 is the total number of handshakes.

Note: It may help to review the solution to Problem 14, where we discussed the solution to a simpler handshaking problem at length.

Solution: There are $7 \times 6 \div 2 = 42 \div 2 = 21$ handshakes among the girls and $5 \times 4 \div 2 = 20 \div 2 = 10$ handshakes among the boys. Add these together to find the total number of handshakes: $21 + 10 = 31$. The solution to this problem can be visualized as two separate complete graphs (see the solution to Problem 14), as shown below. One has 7 vertices and the other has 5 vertices.

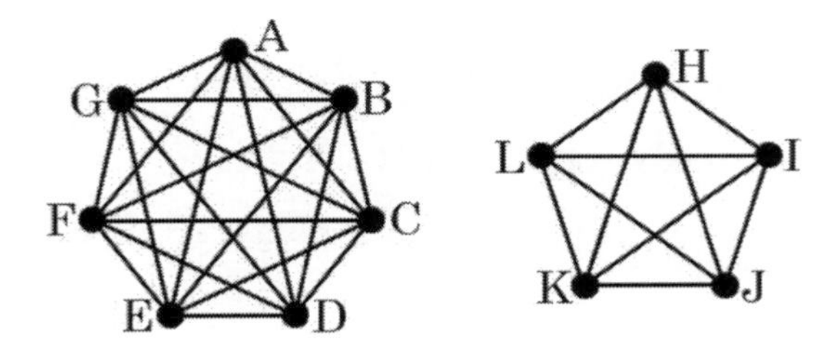

88. Final answer: 35 is the total number of handshakes.

Solution: Each girl shakes hands with all five boys: $7 \times 5 = 35$. This already accounts for all of the handshakes between all 7 girls and 5 boys. This is the final answer.

Note: This solution can be visualized as a complete bipartite graph that has one set of 7 vertices and one set of 5 vertices, where every vertex of the first set connects to every vertex of the second set, but no two vertices from the same set connect to one another, as shown below.

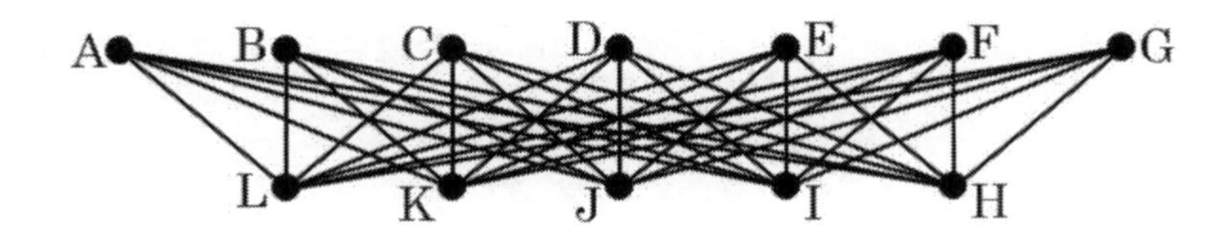

89. Final answers: 20.8 m and 22.8 m are the minimum and maximum perimeter.
Notes: The minimum length and width are $6.4 - 0.2 = 6.2$ and $4.5 - 0.3 = 4.2$. Use these to find the minimum perimeter: $2 \times 6.2 + 2 \times 4.2 = 12.4 + 8.4 = 20.8$. The maximum length and width are $6.4 + 0.2 = 6.6$ and $4.5 + 0.3 = 4.8$. Use these to find the maximum perimeter: $2 \times 6.6 + 2 \times 4.8 = 13.2 + 9.6 = 22.8$.
Note that we could combine the results 20.8 and 22.8 to state that the perimeter equals 21.8 ± 1.0 meters. This problem is concerned with the worst-case error, which is sometimes used in engineering, for example. Note that this is not the only type of error. For example, physicists often propagate errors using a different technique than the worst-case error.

90. Final answers: 26.04 m^2 and 31.68 m^2 are the minimum and maximum area.
Notes: The minimum length and width are $6.4 - 0.2 = 6.2$ and $4.5 - 0.3 = 4.2$. Use these to find the minimum area: $6.2 \times 4.2 = 26.04$ m^2. The maximum length and width are $6.4 + 0.2 = 6.6$ and $4.5 + 0.3 = 4.8$. Use these to find the maximum area: $6.6 \times 4.8 = 31.68$ m^2.
Note that we could combine the results 26.04 and 31.68 to state that the area equals 28.86 ± 2.82 square meters. See the note in the solution to Problem 89 regarding worst-case error.

91. Final answer: The man should travel 5 miles to the east to return along a straight line to his starting position.
Solution: Draw a diagram, like the one below. The dashed line is the answer.

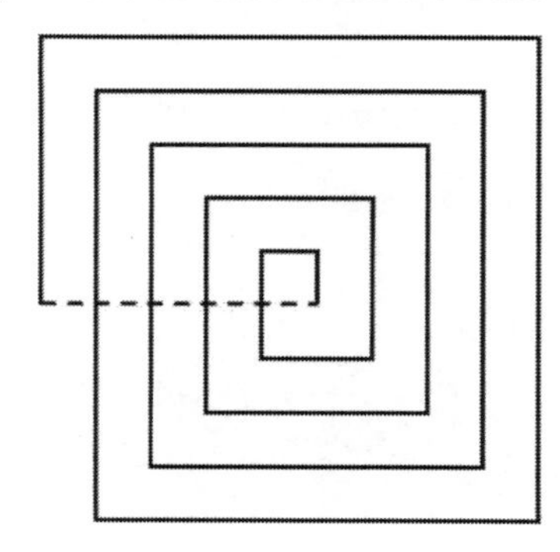

92. Final answer: Greater than 25 percent is the rate of sales needed.

Solution: 800 clicks cost 800 × \$0.75 = \$600.

The company needs to make a profit of greater than \$600 from the products sold through the advertisement in order to make a short-term profit from the advertisement. (If the company also sells products to customers who did not see the advertisement, that is a separate issue. This problem is only concerned with short-term sales generated through the advertisement. To know whether the advertisement is cost-effective, the company should consider short-term sales generated directly from the advertisement as well as potential long-term sales, such as through branding, repeat customers, or word-of-mouth referrals.)

Since the company makes a profit of \$3 per product sold, the company needs to sell more than 200 products through the advertisement, since 200 × \$3 = \$600.

Divide the number of sales needed (more than 200) by the number of clicks (800) and multiply by 100% to determine the needed closing rate as a percent:

$$(200 \div 800) \times 100\% = 1/4 \times 100\% = 100\% \div 4 = 25\%$$

The question specifies that the sales should make a short-term profit. If exactly 200 products were sold, the company breaks even. So the company needs more than 200 sales, or more than 25 percent. (This assumes that the sales resulting from the advertisement are in addition to the sales that the company otherwise makes. These additional sales add to the company's ordinary profits.)

93. Final answer: Meter is what you get when you square the SI units of speed and divide by the SI units of acceleration.

Solution: When we square (m/s), we get m^2/s^2.

Divide m^2/s^2 by the units for acceleration: $(m^2/s^2) \div (m/s^2)$.

To divide by a fraction, multiply by its reciprocal. The reciprocal of m/s^2 is s^2/m.

We get $(m^2/s^2) \div (m/s^2) = (m^2/s^2) \times (s^2/m) = m^2/m = m$.

Note that seconds cancel out because $(1/s^2) \times s^2 = 1$.

Note: This is well-known from physics. For example, if an object travels with uniform acceleration from rest along a straight line, its final speed is given by the formula $v^2 = 2 \times a \times d$, where d is the displacement. Since the whole number 2 does not have any units, this shows that the SI units of speed squared equals the SI units of acceleration times the SI unit of distance.

94. Final answer: 189 batteries can be purchased for $56.

Solution: $56 ÷ $8 = 7. Multiply 27 × 7 = 189.

This problem involves a proportion: 189 is to 27 as $56 is to $8. The two ratios (189:27 and 56:8) both equal 7:1.

95. Final answer: $477.62 will be the balance after six years.

Note: Multiplying the current balance by 1.03 is equivalent to adding 3% to the current balance. For example, 3% of $400 is $12 and 1.03 times $400 equals $412. It is quicker to multiply by 1.03 than it is to multiply by 0.03 and then add.

Solution: Method 1: 1.03 × $400 = $412 (after 1 year)

1.03 × $412 = $424.36 (after 2 years)

1.03 × $424.36 = $437.09 (after 3 years), rounded to the nearest penny

1.03 × $437.09 = $450.20 (after 4 years), rounded to the nearest penny

1.03 × $450.20 = $463.71 (after 5 years), rounded to the nearest penny

1.03 × $463.71 = $477.62 (after 6 years), rounded to the nearest penny

Method 2: $(1.03)^6$ × $400 = 1.19405 × $400 = $477.62 (rounded)

96. Final answer: 144 is the least common multiple (LCM) of 36 and 48.

Check the answer: 36 × 4 = 144 and 48 × 3 = 144.

Notes: Since 36 = 3 × 12 and 48 = 4 × 12, multiply 36 by 4 or multiply 48 by 3 to determine that the least common multiple (LCM) is 144.

97. Final answer: 11 is what the screen displays when the boy types 131.

Solution: Divide the number that the boy types by 4 and find the remainder. The remainder determines what the screen displays.

If the remainder is 0, the screen displays 2. For example, the screen displays 2 for 4 and 8 because 4 ÷ 4 = 1 and 8 ÷ 4 = 2 each have no remainder.

If the remainder is 1, the screen displays 7. For example, the screen displays 7 for 1, 5, and 9 because 1 ÷ 4 = 0R1, 5 ÷ 4 = 1R1, and 9 ÷ 4 = 2R1 each have a remainder of 1.

If the remainder is 2, the screen displays 32. For example, the screen displays 32 for 2, 6, and 10 because 2 ÷ 4 = 0R2, 6 ÷ 4 = 1R2, and 10 ÷ 4 = 2R2 each have a remainder of 2.

If the remainder is 3, the screen displays 11. For example, the screen displays 11 for 3, 7, and 11 because 3 ÷ 4 = 0R3, 7 ÷ 4 = 1R3, and 11 ÷ 4 = 2R3 each have a remainder of 3. Since 131 ÷ 4 = 32R3 (since 4 × 32 = 128 and 131 – 128 = 3), the screen displays 11 when the boy types 131.

98. Final answers: 11/12 and 3/4 (eleven twelfths and three fourths) have a sum of five thirds and a difference of one sixth.

Check the answers: Multiply the numerator and denominator of 3/4 each by 3 to see that 3/4 is equivalent to 9/12. This makes a common denominator of 12.

$$11/12 + 3/4 = 11/12 + 9/12 = 20/12 = 5/3$$

$$11/12 - 3/4 = 11/12 - 9/12 = 2/12 = 1/6$$

Note: Divide 20 and 12 each by 4 to reduce 20/12 to 5/3. Divide 2 and 12 each by 2 to reduce 2/12 to 1/6.

99. Challenge: Our answer, "syrupy," has six letters. Can you beat that?

Notes: Make a chart like the one below to help score the words. Think of words that avoid the letters a-m and which use many letters from t-z.

Letter	a	b	c	d	e	f	g	h	i	j	k	l	m
Value	1	2	3	4	5	6	7	8	9	10	11	12	13

Letter	n	o	p	q	r	s	t	u	v	w	x	y	z
Value	14	15	16	17	18	19	20	21	22	23	24	25	26

Notes: "Syrupy" has a score of (19 + 25 + 18 + 21 + 16 + 25) ÷ 6 = 124 ÷ 6, which equals the repeating decimal 20.66666666... with the 6 repeating forever. This is a 6-letter word with a score higher than 20.

Since the score adds the letter values and divides by the number of letters, the score represents the average letter value.

It is difficult to find a long word with many letters from t-z which does not have a letter from the first half of the alphabet that brings the average letter value below 20. For example, at first "pizzazz" might seem like a great word with 4 z's, but the a and i bring the average letter value below 20.

You should have been able to think of a word with at least five letters that has an average score higher than 20. For example, "fuzzy" is a common word, which has a score of (6 + 21 + 26 + 26 + 25) ÷ 5 = 104 ÷ 5 = 20.8. Another one is "rusty."

100. Final answer: 72 degrees is the angle of each slice from the center.

Solution: Recall from Problem 78 that 360 degrees corresponds to a full circle. Divide 360° by the number of slices: 360° ÷ 5 = 72°.

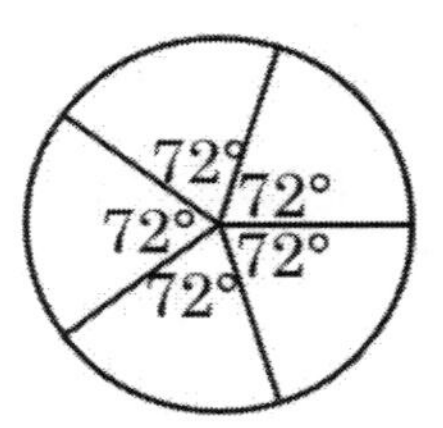

101. Final answer: 12% of the balls are brown when the two bags are mixed.

Try it with numbers: Suppose that the first bag has 100 balls. Since 36% of the balls in the first bag are brown, the first bag will have 36 brown balls and 64 pink balls. The second bag will then have 200 pink balls. When the two bags are mixed, there will be 36 brown balls and a total of 300 balls. Divide 36 by 300 to get 0.12, and multiply by 100% to convert 0.12 to a percent: 0.12 = 12%. Although making up a number for the number of balls in the first bag helps to check the answer with numbers, it is not a desirable way to solve the problem.

A teacher would likely deduct points for solving a problem by making up numbers. One way to solve this problem is to call the number of balls in the first bag N. The number of brown balls in the first bag is then 0.36N. The total number of balls is 3N. Divide 0.36N by 3N to get 0.12, which corresponds to 12%.

102. Final answer: 15 is the greatest common factor (GCF) of 180 and 255.

Check the answer: $180 = 15 \times 12$ and $255 = 15 \times 17$.

Note: 180 factors as $5 \times 3^2 \times 2^2$ and 255 factors as $17 \times 5 \times 3$. Their common factors include one 5 and one 3, which make $5 \times 3 = 15$.

103. Final answer: About 8 minutes is how much time it would take for Nishi and Lee to type a total of one thousand words.

Check the answer: Each minute Nishi and Lee can type a total of $75 + 50 = 125$ words. In 8 minutes, they can type a total of $8 \times 125 = 1000$ words.

104. Final answer: –11 (negative eleven) is the number.

Check the answer: Negative three times negative eleven is positive thirty-three: $-3 \times (-11) = 33$. Two minus signs multiplied together cancel out. Twenty-two minus negative eleven is also thirty-three: $22 - (-11) = 22 + 11 = 33$. Subtracting a negative number equates to addition.

Solution: One way to solve this problem is to apply algebra (though it is not necessary to use algebra to solve this problem):

$$-3n = 22 - n$$

$$-2n = 22$$

$$n = -11$$

In the first step, we added n to both sides. Note that $-3n + n = -2n$. In the next step, we divided both sides by negative two.

105. Final answer: In the fourth February from now, Jean will make her last car payment.

Notes: There are 12 months in one year. The first payment occurs in November.

The 12th payment occurs in October. The 24th and 36th payments are also made in October. The 37th is made in November, the 38th is made in December, the 39th is made in January, and the 40th is made in February.

106. Final answer: 168 is the least common multiple (LCM) of 56 and 84.

Check the answer: $56 \times 3 = 168$ and $84 \times 2 = 168$.

Notes: Since $56 = 2 \times 28$ and $84 = 3 \times 28$, multiply 56 by 3 or multiply 84 by 2 to determine that the least common multiple (LCM) is 168.

107. Final answer: $0.75 per pound is the price per pound of the beans.

Check the answer: The rice costs $\$1.60 \times 3/4 = \$4.80 \div 4 = \$1.20$. The beans cost $\$0.75 \times 2/3 = \$1.50 \div 3 = \$0.50$. The total cost is $\$1.20 + \$0.50 = \$1.70$.

108. Final answers: 48 blocks are in the first stack, 24 blocks are in the second stack, and 12 blocks are in the third stack.

Check the answers: 24 is one-half of 48, 12 is one-half of 24, and $48 + 24 + 12 = 84$ blocks.

Note: One way to solve this problem is to think of the first stack as "one" stack, the second stack as one-half of a stack, and the third stack as one-fourth of a stack. This way, there are 1.75 stacks. The first stack has $84 \div 1.75 = 48$ blocks.

109. Final answer: 1 out of 7 is the probability that the sum will be 9.

Notes: It is instructive to compare this solution to the solution to Problem 27. In Problem 27, the two dice could show the same number, whereas as in this problem, the two discs must have different numbers. The next problem, Problem 110, is more like Problem 27.

Solution: One disc can be any number from 1-7. The second disc can not be the same number as the first disc. One disc has 7 choices, while the other disc has 6 choices. There are a total of $7 \times 6 = 42$ possible outcomes:

$1 + 2 = 3$, $1 + 3 = 4$, $1 + 4 = 5$, $1 + 5 = 6$, $1 + 6 = 7$, $1 + 7 = 8$

$2 + 1 = 3$, $2 + 3 = 5$, $2 + 4 = 6$, $2 + 5 = 7$, $2 + 6 = 8$, $2 + 7 = 9$

$3 + 1 = 4$, $3 + 2 = 5$, $3 + 4 = 7$, $3 + 5 = 8$, $3 + 6 = 9$, $3 + 7 = 10$

$4 + 1 = 5$, $4 + 2 = 6$, $4 + 3 = 7$, $4 + 5 = 9$, $4 + 6 = 10$, $4 + 7 = 11$

$5 + 1 = 6$, $5 + 2 = 7$, $5 + 3 = 8$, $5 + 4 = 9$, $5 + 6 = 11$, $5 + 7 = 12$

$6 + 1 = 7$, $6 + 2 = 8$, $6 + 3 = 9$, $6 + 4 = 10$, $6 + 5 = 11$, $6 + 7 = 13$

$7 + 1 = 8$, $7 + 2 = 9$, $7 + 3 = 10$, $7 + 4 = 11$, $7 + 5 = 12$, $7 + 6 = 13$

There are 6 different ways that the two discs can add up to 9:

$2 + 7 = 9$, $3 + 6 = 9$, $4 + 5 = 9$, $5 + 4 = 9$, $6 + 3 = 9$, $7 + 2 = 9$

The probability that the sum will equal 9 is 6 out of 42, which reduces to 1 out of 7. (Divide 6 and 42 each by 6 to see that 6:42 reduces to 1:7.)

110. Final answer: 6 out of 49 is the probability that the sum will be 9.

Note: The way that this problem is different from the previous problem is that, since the first disc is placed back into the bag, in this problem it is possible for both discs to have the same number.

Solution: There are now $7 \times 7 = 49$ different possible outcomes. We get all 42 outcomes from the solution to Problem 109 plus 7 doubles:

$1 + 1 = 2$, $2 + 2 = 4$, $3 + 3 = 6$, $4 + 4 = 8$, $5 + 5 = 10$, $6 + 6 = 12$, $7 + 7 = 14$

There are still 6 ways for the sum to be 9, just as in Problem 109. The probability that the sum will be 9 is now 6 out of 49 (instead of 6 out of 42).

Note: It is less likely for the sum of the numbers on the two discs to equal 9 in Problem 110 than it was in Problem 109. The reason is that none of the doubles (like $4 + 4 = 8$ or $5 + 5 = 10$) is equal to 9.

111. Final answer: \$51,840 will be the value after three decades.

Note: Multiplying the value by 1.2 is equivalent to adding 20% to the value. For example, 20% of \$30,000 is \$6000 and 1.2 times \$30,000 equals \$36,000. It is quicker to multiply by 1.2 than it is to multiply by 0.2 and then add.

Solution: Method 1: $1.2 \times \$30{,}000 = \$36{,}000$ (after 1 year)

$1.2 \times \$36{,}000 = \$43{,}200$ (after 2 years)

$1.2 \times \$43,200 = \$51,840$ (after 3 years)

Method 2: $(1.2)^3 \times \$30,000 = 1.728 \times \$30,000 = \$51,840$

112. Final answer: 21.6 feet is the diameter of the circle.

Check the answer: The diameter is D = 21.6 ft.

The radius is $R = D \div 2 = 21.6$ ft. $\div 2 = 10.8$ ft.

The circumference is $C = 2 \times \pi \times R = 2 \times \pi \times 10.8 = 21.6 \times \pi$ ft. Since π is 3.14 to two decimal places, $C = 21.6 \times 3.14$ ft. = 67.82 ft. to two decimal places.

The area is $A = \pi \times R^2 = \pi \times 10.8^2$ ft.^{2} = $116.64 \times \pi$ ft.^{2} Since π is 3.14 to two decimal places, $A = 116.64 \times 3.14$ ft.^{2} = 366.25 ft.^{2} to two decimal places.

Observe that $A \div C = 366.25$ ft.^{2} $\div$ 67.82 ft. = 5.4 ft.

Notes: It is not necessary to use the numerical value of π. It will cancel out:

$$A \div C = (116.64 \times \pi \text{ ft.}^2) \div (21.6 \times \pi \text{ ft.}) = 5.4 \text{ ft.}$$

One way to solve this problem is:

$$A \div C = (\pi \times R^2) \div (2 \times \pi \times R) = R \div 2 = 5.4 \text{ ft.}$$

$$R = 2 \times 5.4 \text{ ft.} = 10.8 \text{ ft.}$$

113. Final answers: Uma is 13 years old, Xander is 40 years old, and Zoe is 63 years old.

Check the answers: $40 - 13 = 27$, $63 - 40 = 23$, and $13 + 40 + 63 = 116$.

114. Final answer: 40 is the number.

Check the answer: Zero minus 40 equals negative 40. That is, $0 - 40 = -40$.

One-half of forty is 20. Adding 20 to negative 60 equals negative 40. That is,

$$40/2 + (-60) = 20 - 60 = -40$$

Solution: One way to solve this problem is to apply algebra (though it is not necessary to use algebra to solve this problem):

$$0 - n = n/2 + (-60)$$

$$-n = n/2 - 60$$

$$60 = 3n/2$$

$$2 \times 60 / 3 = n$$

$$40 = n$$

After the second step, we added n to both sides and we also added 60 to both sides. Note that n + n/2 = 3n/2 because 1 plus 1/2 = 3/2. After the next step, we multiplied both sides by 2/3.

115. Final answer: 62.5 meters is the actual height of the building.

Check the answer: 3 centimeters corresponds to 25 meters. Since 7.5 ÷ 3 = 2.5, we need to multiply 25 meters by 2.5. The answer is 25 × 2.5 = 62.5 meters. This problem involves a proportion: 62.5 is to 25 as 7.5 is to 3. The two ratios (62.5:25 and 7.5:3) both equate to 2.5:1 (which is equivalent to 5:2).

116. Final answer: 84 is the two-digit number.

Check the answer: The sum of the digits of 84 is 8 + 4 = 12. When the sum of the digits is subtracted from 84, we get 84 – 12 = 72. Also, the tens digit (8) is twice the units digit (4).

117. Final answer: 120 pixels is how much farther the yellow dot has traveled compared to the purple dot when the two dots meet.

Note: Recall the following rate equation from the solutions to Problems 10 and 31:

$$\text{distance} = \text{speed} \times \text{time}$$

Check the answer: The two dots meet in 60 seconds. The yellow dot travels a distance equal to 7 × 60 = 420 pixels. The purple dot travels a distance equal to 5 × 60 = 300 pixels. The total distance traveled by the two dots is 420 + 300 = 720 pixels. The yellow dot travels 420 – 300 = 120 pixels farther.

118. Final answer: 585 is the sum of the multiples of 9 between 9 and 100.

Notes: 18 + 81 = 27 + 72 = 36 + 63 = 45 + 54 = 99. The answer is

$$99 \times 4 + 90 + 99 = 99 \times 5 + 90 = 495 + 90 = 585$$

119. Final answer: 120 is the total number of origami sheets.

Check the answers: Initially, Connie had 12, Drew had 36, and Emily had 72.

Initially, Emily had twice as many origami sheets as Drew: 72 = 2 × 36.

Initially, Drew had three times as many origami sheets as Connie: 36 = 3 × 12.

Emily gave 28 sheets to Connie and 4 sheets to Drew.

Emily now has 72 – 28 – 4 = 40 sheets.

Drew now has 36 + 4 = 40 sheets.

Connie now has 12 + 28 = 40 sheets.

All three now have 40 sheets. The total number of sheets is 40 + 40 + 40 = 120.

120. Final answer: 0.01 (one hundredth) becomes 0.000001 (one millionth) when it is cubed. The answer (0.01) is equivalent to 1/100.

Check the answer: $0.01^3 = 0.01 \times 0.01 \times 0.01 = 0.000001$.

In fractional form, $(1/100)^3 = (1/100) \times (1/100) \times (1/100) = 1/1{,}000{,}000$.

121. Final answer: 8 hours and 45 minutes is how much she has slept per day, on average.

Check the answer: Since there are 60 minutes in one hour, 45 minutes is 3/4 of one hour, which equates to 0.75 hours. Thus, 8 hours and 45 minutes equates to 8.75 hours. Multiply 8.75 hours by 7 to find that the woman slept for a total of 61.25 hours (which is 61 hours and 15 minutes). The total number of hours in one week is 7 × 24 = 168. Subtract 61.25 from 168 to get 106.75 hours, which is the number of hours that the woman has been awake. This equates to 106 hours and 45 minutes.

Note: In case you were wondering about the wording of the problem regarding midnight, the problem is trying to emphasize that the week in question consists of exactly seven 24-hour periods. This makes the total time exactly 168 hours. It does not matter whether or not the woman had already been sleeping when the week began. Only the total time spent sleeping during the week matters.

122. Final answer: 16 seeds are in each of the first 7 rows, and 12 seeds are in the 8th row.

Check the answer: $16 \times 7 + 12 = 112 + 12 = 124$

Note: If you think of it as 7 complete rows plus 0.75 rows, one way to solve the problem is $124 \div 7.75 = 16$.

123. Final answers: $111 \times 222 = 110{,}112$ with all three numbers written in base 3. If all three numbers are written in base 10, it is $13 \times 26 = 338$.

Notes: The first 27 numbers are written in base 10 and base 3 below.

Base 10	1	2	3	4	5	6	7	8	9
Base 3	1	2	10	11	12	20	21	22	100

Base 10	10	11	12	13	14	15	16	17	18
Base 3	101	102	110	111	112	120	121	122	200

Base 10	19	20	21	22	23	24	25	26	27
Base 3	201	202	210	211	212	220	221	222	1000

We thus see that 111×222 in base 3 is equivalent to 13×26 in base 10.

In base 10, it is easy to determine that $13 \times 26 = 338$. Now we need to convert 338 from base 10 to base 3 to determine the answer to 111×222 in base 3.

Note the following helpful conversions between base 10 and base 3:

Base 10	$3^1 = 3$	$3^2 = 9$	$3^3 = 27$	$3^4 = 81$	$3^5 = 243$
Base 3	10	100	1000	10,000	100,000

In base 10, we can rewrite 338 as $338 = 243 + 81 + 14$. This shows that 338 in base 10 is equivalent to $100{,}000 + 10{,}000 + 112 = 110{,}112$ in base 3.

124. Final answers: The original rectangle has a length of 24 units and a width of 16 units.

Check the answers: The perimeter of the original rectangle is $2 \times 24 + 2 \times 16 = 48 + 32 = 80$ units. The new rectangle has a length of $2 \times 24 = 48$ units. The perimeter of the new rectangle is $2 \times 48 + 2 \times 16 = 96 + 32 = 128$ units.

125. Final answer: 780 is the least common multiple (LCM) of 12, 20, and 52.

Check the answer: $12 \times 65 = 780$, $20 \times 39 = 780$, and $52 \times 15 = 780$.

Notes: Since $12 = 3 \times 4$, $20 = 5 \times 4$, and $52 = 13 \times 4$, multiply 12 by $5 \times 13 = 65$, multiply 20 by $3 \times 13 = 39$, or multiply 52 by $3 \times 5 = 15$ to determine that the least common multiple (LCM) is 780.

126. Final answer: 264 is the three-digit number.

Check the answer: $462 - 264 = 198$, the second digit (6) is 2 more than the third digit (4), and the sum of the digits is $2 + 6 + 4 = 12$.

127. Final answer: 13 out of 256 is the probability of scoring at least 75%.

Notes: Each question can be answered A, B, C, or D. There are $4 \times 4 \times 4 \times 4 =$ 256 different ways of selecting one answer to each of the four questions. You could answer AAAA, AAAB, AAAC, AAAD, AABA, AABB, AABC, AABD, AACA, and so on, all of the way up to DDDD.

One of the 256 possibilities has 4 correct answers, earning a score of 100%.

There are 12 ways to have 3 correct answers, earning a score of exactly 75%. There are 3 ways to answer Question 1 incorrectly and Questions 2-4 correctly because Question 3 has 1 correct answer and 3 incorrect answers. Similarly, there are 3 ways to answer Question 2 incorrectly and Questions 1, 3, and 4 correctly, there are 3 ways to answer Question 3 incorrectly and Questions 1, 2, and 4 correctly, and there are 3 ways to answer Question 4 incorrectly and Questions 1-3 correctly.

The 1 way to score 100% plus the 12 ways to score exactly 75% gives a total of $12 + 1 = 13$ ways to score at least 75%. The likelihood of scoring at least 75% is 13 out of 256 (which is approximately 5%).

128. Final answer: Second is what you get when you divide the SI unit of distance by the SI units of acceleration and then take the square root of that.

Solution: The SI unit of distance is m and the SI units of acceleration are m/s^2.

Divide m by the units for acceleration: m ÷ (m/s^2)

To divide by a fraction, multiply by its reciprocal. The reciprocal of m/s^2 is s^2/m.

We get m ÷ (m/s^2) = m × (s^2/m) = m × s^2 / m = s^2.

Note that meters cancel out because m / m = m ÷ m = 1.

So far, what we have is m ÷ (m/s^2) = s^2.

The problem states that we now need to take the square root. The square root of s^2 is s. The answer is "seconds."

Note: This is well-known from physics. If an object is dropped from rest in a uniform gravitational field, the time of descent is given by:

$$t = \sqrt{\frac{2h}{g}}$$

Since the whole number 2 does not have any units, this shows that the SI units of distance divided by the SI units of gravitational acceleration equals a second squared, such that after taking the square root the time will be in seconds.

129. Final answer: \$4.49 or less is the price that brand B's 12-pack must be in order for it to be more economical than brand A's 8-pack.

Check the answer: Brand A's 8-pack provides 8 × 60 = 480 hours of battery life. At a cost of \$4, Brand A's 8-pack provides 480 ÷ 4 = 120 hours per dollar spent. Brand B's 12-pack provides 12 × 45 = 540 hours of battery life. (Although each battery does not last as long, by having more batteries in one pack, the pack provides more hours overall.) If Brand B's 12-pack is priced at \$4.50, Brand B's 12-pack would provide 540 ÷ 4.50 = 120 hours per dollar spent. However, this price would make Brand B's 12-pack equally economical as Brand A's 8-pack, but the problem asked for Brand B's 12-pack to be more economical. So if the price is \$4.49 or lower, Brand B's 12-pack will be more economical.

130. Final answer: 512 cubic units is the volume of the cube.

Check the answer: The edge length of the cube is 8 units.

The volume of the cube is $8^3 = 8 \times 8 \times 8 = 512$ cubic units.

Since the cube has 6 square sides, the total surface area of the cube is $6 \times 8^2 = 6 \times 64 = 384$ square units.

131. Final answers: 0.74 and 0.46 have a sum of 1.2 and a difference of 0.28.

Check the answers: $0.74 + 0.46 = 1.2$ and $0.74 - 0.46 = 0.28$.

132. Final answer: Fernando and Gilbert will meet on step 320.

Check the answers: Fernando will climb $8 \times 40 = 320$ stairs in the time that Gilbert descends $17 \times 40 = 680$ stairs. Since $1000 - 680 = 320$, Fernando and Gilbert will meet on step 320.

Notes: Fernando basically begins at step 0. His first step takes him to step 1. One way to solve this problem is to call the time it takes for Fernando to climb 8 stairs one time period. In one time period, Fernando and Gilbert get $8 + 17 = 25$ stairs closer. It will take them $1000 \div 25 = 40$ time periods to climb a total of 1000 stairs. In 40 time periods, Fernando climbs $8 \times 40 = 320$ stairs while Gilbert descends $17 \times 40 = 680$ stairs.

133. Final answer: 100 plus 100 squared plus 100 cubed equals 1,010,100.

Check the answer: $100 + 100^2 + 100^3 = 100 + 10{,}000 + 1{,}000{,}000 = 1{,}010{,}100$.

Notes: The trick is to write 1,010,100 in expanded form:

$$1{,}010{,}100 = 1{,}000{,}000 + 10{,}000 + 100$$

Now note that $100^2 = 100 \times 100 = 10{,}000$ and $100^3 = 100 \times 100 \times 100 = 1{,}000{,}000$.

134. Final answer: 37.5% is the minimum percent discount that Marco needs.

Check the answer: The printer costs \$200. A 37.5% discount is $\$200 \times 0.375 = \$75$. This will bring the price down to $\$200 - \$75 = \$125$. The tax rate is 8%. The tax is $0.08 \times \$125 = \$10$. The total cost, including tax, is $\$125 + \$10 = \$135$.

Note: One way to solve this problem is to first divide $\$135 \div 1.08 = \$125$. This shows that the cost before tax needs to be \$125. The discount needs to be at least $\$200 - \$125 = \$75$, which corresponds to $\$75 \div \$200 \times 100\% = 37.5\%$.

135. Final answer: 18 is the greatest common factor (GCF) of 108 and 450.

Check the answer: $108 = 6 \times 18$ and $450 = 25 \times 18$.

Note: 108 factors as $3^3 \times 2^2$ and 450 factors as $5^2 \times 3^2 \times 2$. Their common factors include 3^2 and one 2, which make $3^2 \times 2 = 9 \times 2 = 18$.

136. Final answers: 8 out of 27 is the probability of 3 unfair coins all landing heads up, while 1 out of 8 is the probability of 3 fair coins all landing heads up.

Solution: We will address the **fair** coins first. There are 8 possible outcomes. Note that H = heads and T = tails.

HHH, HHT, HTH, THH, HTT, THT, TTH, TTT

If the coins are fair, then H and T are equally likely. The likelihood that the three coins will all be heads (HHH) is 1 out of 8. Another way to determine this is to realize that the probability of each coin being heads is 1/2:

$$(1/2)^3 = 1/2 \times 1/2 \times 1/2 = 1/8$$

This is the key to dealing with the **unfair** coins. Replace 1/2 with 2/3:

$$(2/3)^3 = 2/3 \times 2/3 \times 2/3 = 8/27$$

137. Final answer: A maximum of 9 such line segments can be drawn.

Notes: This diagram is called a maximal planar graph. Specifically, this is the pentahedral graph, which has 5 vertices and 9 edges, as shown below. It has one less edge than a complete graph with 5 vertices (see the right diagram in the solution to Problem 87). Whereas a complete graph with 5 vertices can only be drawn by having at least one pair of edges cross, a maximal planar graph can be drawn in the plane without any crossings (as shown below). With 10 edges, there would be at least one crossing. (Of course, edges may meet at a vertex.)

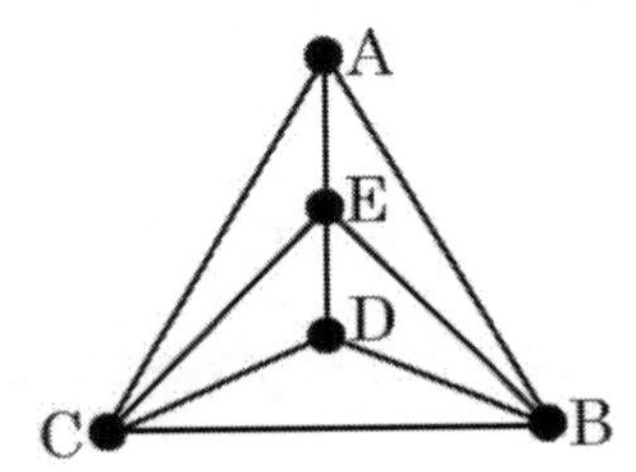

138. Final answers: 6 cans of paint are needed. The excess paint can cover an additional 92 square feet.

Check the answers: Although 1 yard equates to 3 feet, note that 1 square yard equates to 9 square feet. Why? One square yard is a square where each side is 1 yard. Each side of the square is thus 3 feet. If a square has an edge length of 3 feet, its area is 3 × 3 = 9 square feet. You can see this visually below.

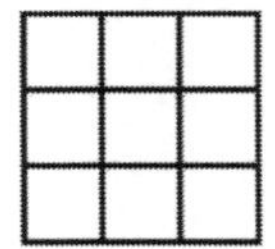

Therefore, 48 square yards equates to 48 × 9 = 432 square feet.

6 cans of paint can cover 6 × 432 = 2592 square feet. An additional 2592 – 2500 = 92 square feet can be painted with the leftover paint.

139. Final answer: The number could be 8 or –8. (The answer is ±8.)

Check the answer: ±8 times 3 equals ±24. Square this to get (±24) × (±24) = 576. Add two dozen to get 576 + (2 × 12) = 576 + 24 = 600. Note that –24 × –24 = 576.

140. Final answer: 13 to 11 is the ratio of washers to bolts in the barrel.

Try it with numbers: Suppose that the sum of the washers and bolts is equal to 60 in each barrel. (The numbers work out nicely in this problem if the sum is a multiple of 12.) The first bucket has 45 washers and 15 bolts. It has 3 times as many washers as bolts, and 45 + 15 = 60. The second bucket has 20 washers and 40 bolts. It has one-half as many washers as bolts, and 20 + 40 = 60. When the two buckets are poured into the same barrel, the barrel has 45 + 20 = 65 washers and 15 + 40 = 55 bolts. The ratio of washers to bolts for the barrel is 65:55, which reduces to 13:11 if you divide 65 and 55 each by 5. Although making up a number for the sum of the washers and bolts in a bucket helps to check the answer with numbers, it is not a desirable way to solve the problem. A teacher would likely deduct points for solving a problem by making up numbers. One way to solve this problem is to call the number of bolts in the first bucket N. The number of washers in the first bucket is 3N. The sum of the washers and bolts is 4N. The number of bolts in the second bucket plus half as many washers also equals 4N. The number of bolts in the second bucket must be 8N/3. The number of washers in the second bucket is 4N/3. Note that 8N/3 + 4N/3 = 12N/3 = 4N. The number of washers in the barrel is 3N + 4N/3 = 9N/3 + 4N/3 = 13N/3. The number of bolts in the barrel is N + 8N/3 = 3N/3 + 8N/3 = 11N/3. The ratio of washers to bolts in the barrel is thus 13:11.

141. Final answer: 500 is how many hundred-dollar bills you could exchange one million nickels for.

Check the answer: 1,000,000 nickels has a value of 1,000,000 × \$0.05 = \$50,000. 500 hundred-dollar bills is also worth 500 × \$100 = \$50,000.

Note: Another way to see that a million nickels is worth \$50,000 is to note that one nickel is 1/20 of a dollar: \$1,000,000 ÷ 20 = \$50,000.

142. Final answer: 32 passengers are still aboard the train.

Solution: Method 1: 243 ÷ 3 = 81 and 243 – 81 = 162 (after 1 stop)

162 ÷ 3 = 54 and 162 – 54 = 108 (after 2 stops)

108 ÷ 3 = 36 and 108 – 36 = 72 (after 3 stops)

72 ÷ 3 = 24 and 72 – 24 = 48 (after 4 stops)

48 ÷ 3 = 16 and 48 – 16 = 32 (after 5 stops)

Method 2: $243 \times (2/3)^5 = 243 \times 2^5 \div 3^5 = 243 \times 32 \div 243 = 32$

143. Final answer: 13/36 is the average of 1/2, 1/3, and 1/4.

Solution: (1/2 + 1/3 + 1/4) ÷ 3 = (6/12 + 4/12 + 3/12) ÷ 3 = (13/12) ÷ 3

Note that 3 = 3/1. To divide by a fraction, multiply by its reciprocal:

(13/12) ÷ (3/1) = (13/12) × (1/3) = 13/36

144. Final answer: 40 seconds is how much time it takes for Katherine to pass Larry by exactly one lap.

Note: Recall the following rate equation from the solutions to Problems 10, 31, and 117:

$$\text{distance} = \text{speed} \times \text{time}$$

Check the answer: In 40 seconds, Katherine travels $24 \times 40 = 960$ meters and Larry travels $18 \times 40 = 720$ meters. In 40 seconds, Katherine travels $960 - 720 = 240$ meters farther than Larry, which is exactly one lap.

Notes: One way to solve this problem is to note that Katherine travels $24 - 18 = 6$ meters farther than Larry each second. It will take $240 \div 6 = 40$ seconds for Katherine to be exactly one circumference past Larry.

(We do not need to use the formula for circumference because the problem gave us the circumference. We do not need to use diameter or radius. If the problem had given diameter or radius instead of circumference, then we would have used the formula to find the circumference.)

145. Final answers: Isabel is 12 years old now and Veronica is 24 years old now.

Check the answers: Veronica is twice as old as Isabel: $24 = 2 \times 12$.

The problem also states that Veronica's current age (which is 24) is three times what Isabel's age was four years ago (which was $12 - 4 = 8$): $24 = 3 \times 8$.

146. Final answers: Each row in the right section has 25 seats and each row in the left section has 15 seats.

Check the answers: The right section has $25 \times 50 = 1250$ seats while the left section has $15 \times 30 = 450$ seats. The total number of seats is $1250 + 450 = 1700$. 40% of 25 is equal to $0.4 \times 25 = 10$. Subtract 10 from 25 to get 15. That is, 15 is 40% less than 25.

Notes: One way to solve this problem is to think of each row of the right section as a unit and each row of the left section as 0.6 of a unit (since it is 40% shorter). There are then $50 + 0.6 \times 30 = 50 + 18 = 68$ units. The number of seats in a row of the right section is $1700 \div 68 = 25$.

147. Final answer: 31 is the sum of the prime factors of 1190.

Check the answer: The prime factorization of 1190 is

$$1190 = 17 \times 7 \times 5 \times 2$$

The sum of the prime factors is $17 + 7 + 5 + 2 = 31$.

148. Final answer: 1 out of 22 is the probability that both blocks will be green.

Solution: There are initially $5 + 4 + 3 = 12$ blocks. Initially, there are 3 green blocks. When the first block is selected, the likelihood of selecting one green block is 3 out of 12, which reduces to 1 out of 4 (if you divide 3 and 12 each by 3). When the second block is selected, there are 11 blocks to choose from. If the first block was green (which is necessary if both blocks are to be green), there will be 2 green blocks available when the second block is selected. The likelihood of selecting one green block when there are 2 green blocks and 11 blocks total is 2 out of 11. To combine the two results together, multiply the probabilities. Multiply 1/4 for the first block to be green times 2/11 for the second block to also be green: $(1/4) \times (2/11) = 2/44 = 1/22$.

149. Final answer: 0.54 (which is equivalent to 27/50) is 36% of 5/8 of 2.4.

Solution: 5/8 of 2.4 equals $5/8 \times 2.4 = 12/8 = 12 \div 8 = 1.5$.

36% of 1.5 equals $0.36 \times 1.5 = 0.54$ (which is equivalent to 27/50).

150. Final answers: Angie has \$22.80 and Steve has \$15.60.

Check the answers: Angie has 72 quarters and 48 dimes. Angie's coins have a total value of

$$72 \times \$0.25 + 48 \times \$0.10 = \$18 + \$4.80 = \$22.80$$

Steve has 24 quarters and 96 dimes. Steve's coins have a total value of

$$24 \times \$0.25 + 96 \times \$0.10 = \$6 + \$9.60 = \$15.60$$

Angie has three times as many quarters as Steve: $72 = 3 \times 24$.

Angie has half as many dimes as Steve: $48 = 0.5 \times 96$.

Together, Angie and Steve have $72 + 24 = 96$ quarters and $48 + 96 = 144$ dimes.

Since 50% of 96 is 48 and since 96 + 48 = 144, the total number of dimes (144) is 50% more than the total number of quarters (96).

The total value of their money is \$22.80 + \$15.60 = \$38.40.

151. Final answers: 32 small gloves, 72 medium gloves, and 40 large gloves.

Check the answers: 32:72:40 reduces to 4:9:5 if you divide 32, 72, and 40 each by 8. The total number of gloves is 32 + 72 + 40 = 144.

Notes: One way to solve the problem is to divide 144 by (4 + 9 + 5). Then multiply this by 4, by 9, and by 5 to get the three answers.

152. Final answer: 100,000 (one hundred thousand) centimeters are in 1 km.

Check the answer: One centimeter is 0.01 meters (or 1/100 of a meter).

One kilometer is 1000 meters. Note that 1000 ÷ 0.01 = 100,000.

153. Final answer: 7/16 (seven sixteenths) remained after everyone ate.

Solution: Method 1: 1 – 1/8 = 8/8 – 1/8 = 7/8 remained after the daughter ate.

7/8 – 7/8 × 1/6 = 7/8 – 7/48 = 42/48 – 7/48 = 35/48 remained after the son ate.

35/48 – 35/48 × 1/5 = 35/48 – 7/48 = 28/48 = 7/12 remained after the mother ate.

7/12 – 7/12 × 1/4 = 7/12 – 7/48 = 28/48 – 7/48 = 21/48 = 7/16 remained after the father ate. Note that 21/48 reduces to 7/16 if you divide 21 and 48 each by 3.

Method 2: (7/8) × (5/6) × (4/5) × (3/4) = 7 × 5 × 4 × 3 ÷ 8 ÷ 6 ÷ 5 ÷ 4

= 7 × 3 ÷ 8 ÷ 6 = 21 ÷ 48 = 7/16

Note that 5 × 4 ÷ 5 ÷ 4 = 1. The arithmetic is simpler if you cancel the 5 and 4.

154. Final answer: 40 cm will be the distance when the force is 36 N.

Notes: The force in Newtons times the distance in cm always equals 1440:

2 N × 720 cm = 1440 N cm

5 N × 288 cm = 1440 N cm

9 N × 160 cm = 1440 N cm

36 N × 40 cm = 1440 N cm

It is not necessary to know any science to answer this question.

155. Final answer: 1.5 seconds is how much time it takes the bug to traverse the full length of the truck (relative to the truck).

Note: Recall the following rate equation from the solutions to Problems 10, 31, 117, and 144:

$$\text{distance} = \text{speed} \times \text{time}$$

Check the answer: In 1.5 seconds, the truck travels $18 \times 1.5 = 27$ m relative to the ground while the bug travels $26 \times 1.5 = 39$ m relative to the ground. During this time, the bug travels $39 - 27 = 12$ m farther than the truck, which equals the length of the truck.

Note: One way to solve this problem is to realize that the speed of the bug is 8 m/s faster than the truck, so it will take $12 \div 8$ s for the bug to pass the truck.

156. Final answer: 3/11 (three elevenths) of a second, which is 0.27272727… (with the 27 repeating forever) in decimal form, is how much time would pass.

Check the answer: In 3/11 seconds, the truck travels $18 \times 3/11 = 54/11$ m in one direction while the bug travels $26 \times 3/11 = 78/11$ m in the opposite direction. Note that $54/11 + 78/11 = 132/11 = 12$ meters.

Notes: This time the relative speed is $26 + 18 = 44$ m/s. We add the speeds in this problem because the truck and bug travel in opposite directions (whereas we subtracted the speeds in the previous problem where the direction was the same). Here is another way to look at it. Suppose one car travels 75 mph while another travels 70 mph. If both cars travel in the same direction and collide, the relative speed is $75 - 70 = 5$ mph, whereas if the cars travel in the opposite direction and collide, the relative speed is $75 + 70 = 155$ mph. A head-on collision is far worse because the speeds add to make the relative speed. Back to the truck and bug, the answer is $12 \div 44 = 3/11$ seconds.

157. Final answer: 135 square feet is the actual area of the room.

Check the answer: 0.25 inches corresponds to 3 feet, so 0.75 inches corresponds to 9 feet and 1.25 inches corresponds to 15 feet.

The actual area of the room is 9 feet × 15 feet = 135 square feet.

Note: If you first determine that the area of the scaled room is 0.75 in. × 1.25 in. = 0.9375 square inches, you will need to multiply by 12 × 12 = 144 to convert from scaled square inches to actual square feet. (Although 1 real foot = 12 real inches, here we are converting from scaled inches to real feet, which is different. Also, we're converting with square units.) Why is it 12 × 12 instead of just 12? Review the solution to Problem 138 to understand this. Anyway, it is simpler to convert to actual feet before finding the area.

158. Final answers: 46% will be red, 34% will be blue, and 20% will be yellow.

Solution: Initially, 0.34 × 50 = 17 regions are red, 0.42 × 50 = 21 regions are blue, and 0.24 × 50 = 12 regions are yellow. (Note that 17 + 21 + 12 = 50.)

If 4 blue regions are changed to red, this makes 21 red, 17 blue, and 12 yellow. If 2 yellow regions are also changed to red, this makes 23 red, 17 blue, and 10 yellow. (Note that 23 + 17 + 10 = 50.)

The final percents are 23 ÷ 50 × 100% = 46% red, 17 ÷ 50 × 100% = 34% blue, and 10 ÷ 50 × 100% = 20% yellow. (Note that 46% + 34% + 20% = 100%.)

159. Final answer: 33 people are at the family get-together.

Solution: 1 woman + 2 twins + 6 triplets + 24 quadruplets = 33 people

Note: There are 2 × 3 = 6 triplets because each twin had 3 children. There are 6 × 4 = 24 quadruplets because each triplet had 4 children.

160. Final answer: 120 different 6-digit numbers have one 4, one 5, three 7's, and one 9.

Note: Recall the permutation formula N! ÷ M! from the solution to Problem 20, where N! means to multiply N times (N – 1) times (N – 2) and so on until reaching one. Here, N = 6 is the number of digits and M = 3 because the digit 7 appears 3 times in each number.

With the formula, we get 6! ÷ 3! = (6 × 5 × 4 × 3 × 2 × 1) ÷ (3 × 2 × 1) = 720 ÷ 6 = 120.

161. Final answer: Meters per second is what you get when you multiply the SI unit of distance by the SI units of acceleration and then take the square root of that.

Solution: The SI unit of distance is m and the SI units of acceleration are m/s^2.
Multiply m by the units for acceleration: m × (m/s^2) = m^2/s^2.
The problem states that we now need to take the square root. The square root of m^2/s^2 is m/s. The answer is "meters per second."
Note: This is well-known from physics. If an object uniformly accelerates along a straight line starting from rest, the final speed is given by:

$$v = \sqrt{2ad}$$

Since the whole number 2 does not have any units, this shows that the SI unit of distance times the SI units of acceleration equals m^2/s^2, such that after taking the square root the final speed will be in meters per second.

162. Final answer: 1/16 (one sixteenth) is the fraction of the area that remains.
Draw it: Each time the sheet is cut so that the side that had been shorter (prior to cutting the paper) remains unchanged. The original paper is 8.5 in. by 11 in. For the first cut (second figure from the left), the 8.5-in. side remains 8.5 in. while the 11-in. side becomes 11 ÷ 2 = 5.5 in. For the second cut (middle figure), the 5.5-in side remains 5.5 in. (because it is smaller than the 8.5-in. side) while the 8.5-in. side becomes 8.5 ÷ 2 = 4.25 in. For the third cut (second figure from

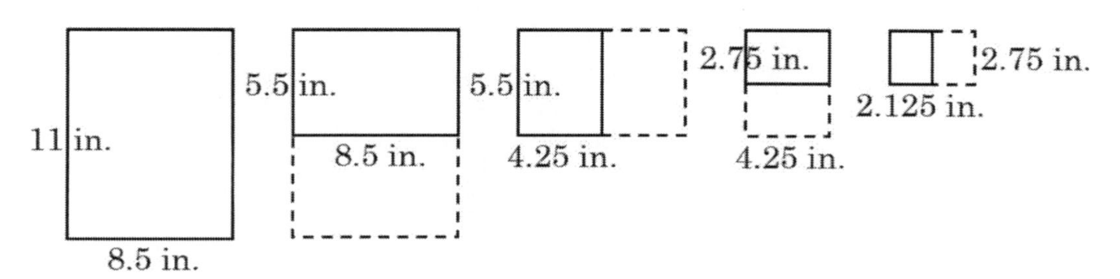

the right), the 4.25-in. side remains 4.25 in. while the 5.5-in. side becomes 5.5 ÷ 2 = 2.75 in. For the final cut (right figure), the 2.75-in. side remains 2.75 in. while the 4.25-in. side becomes 4.25 ÷ 2 = 2.125 in.

Check the answer: The final area is 2.125 in. × 2.75 in. = 5.84375 square inches. The initial area was 8.5 in. × 11 in. = 93.5 square inches. Find the ratio of the areas: 5.84375 ÷ 93.5 = 0.0625 = 1/16.

Solution: The solution is very simple because the area is cut in half four times:

$$(1/2)^4 = (1/2) \times (1/2) \times (1/2) \times (1/2) = 1/16$$

Note: You can draw the diagram differently and still obtain the correct answer, provided that each sheet has one-half of the area of the previous sheet.

163. Final answer: 56.25 is the number.

Check the answer: Note that $7.5^2 = 7.5 \times 7.5 = 56.25$.

This means that the square root of 56.25 equals 7.5.

$$20 - \sqrt{56.25} = 20 - 7.5 = 12.5$$
$$5 + \sqrt{56.25} = 5 + 7.5 = 12.5$$

Notes: Imagine that the problem had said, "Twenty minus a number equals five plus the same number." One way to solve this (different) problem would be to write 20 – N = 5 + N (though it is not necessary to use algebra), which leads to N = 7.5. Now let N equal the square root of another number, M. Square both sides to see that $N^2 = M$, which shows that $M = 7.5^2 = 56.25$.

164. Final answers: Gina has traveled 625 m and Hank has traveled 375 m.

Note: Recall the following rate equation from the solutions to Problems 10, 31, 117, 144, and 155:

$$\text{distance} = \text{speed} \times \text{time}$$

Check the answers: The time is 125 seconds.

Gina has traveled 5 × 125 = 625 m. Hank has traveled 3 × 125 = 375 m.

Observe that 625 m + 375 m = 1000 m.

Notes: One way to solve the problem is to realize that Gina and Hank travel a total of 8 meters each second. Note that one kilometer equals 1000 meters. The time is $1000 \div 8 = 125$ s. Gina travels $5 \times 125 = 625$ m and Hank travels $3 \times 125 = 375$ m. Another way to solve the problem is to note that Gina travels 5/8 of one kilometer while Hank travels 3/8 of one kilometer.

165. Final answers: \$1.80 is the price of a bottle of water and \$3.50 is the price of a bowl of fruit.

Check the answers: 5 × \$1.80 + 3 × \$3.50 = \$9 + \$10.50 = \$19.50

4 × \$1.80 + 7 × \$3.50 = \$7.20 + \$24.50 = \$31.70

166. Final answer: 48 centimeters is the perimeter.

Check the answers: The length is 18 cm and the width is 6 cm. Note that $18 = 3 \times 6$. The area is $18 \times 6 = 108$ cm^2. The perimeter is:

$$2 \times 18 + 2 \times 6 = 36 + 12 = 48 \text{ cm}$$

167. Final answer: 48 minutes is about how much time it will take for Jordan and Kim to finish the jigsaw puzzle if they work together.

Notes: 2 hours equates to 120 minutes. 1 hour and 20 min. equates to 80 min.

Solution: Method 1: Suppose that there are N pieces in the puzzle.

Since Jordan can complete the puzzle in 120 min., Jordan puts together N/120 pieces per min., on average. Since Kim can complete the puzzle in 80 min., Kim puts together N/80 pieces per min., on average.

If Jordan and Kim work together, they will put together N/120 + N/80 pieces per min., on average. (This assumes that they work just as efficiently together as they do alone, and they do not interfere with each other while solving the puzzle. Since the problem did not mention such interference, it is a necessary assumption.) Make a common denominator: 2N/240 + 3N/240 = 5N/240 = N/48. Jordan and Kim put together N/48 pieces per min., on average, when they work together. It will thus take them about 48 minutes to complete the puzzle.

Method 2: Add the times in reciprocal, and then take the reciprocal of the answer (which is effectively what we did in Method 1): 1/120 + 1/80 = 2/240 + 3/240 = 5/240 = 1/48. The reciprocal of 1/48 is 48. The answer is 48 minutes.

Notes: It should make sense that Jordan and Kim complete the puzzle in less time (48 minutes) when they work together than if either worked alone. Note that 48 minutes is equivalent to 0.8 hours.

168. Final answer: 42 is the greatest common factor (GCF) of 210, 336, and 504.

Check the answer: 210 = 5 × 42, 336 = 8 × 42, and 504 = 12 × 42.

Note: 210 factors as 7 × 5 × 3 × 2, 336 factors as $7 \times 3 \times 2^4$, and 504 factors as $7 \times 3^2 \times 2^3$. Their common factors include 7, 3, and 2, which make 7 × 3 × 2 = 42.

169. Final answer: 1 foot and 2 inches is the length of each piece.

Solution: 8 + 1/3 – 2.5 = 5.5 + 1/3 = 5 + 1/2 + 1/3 = 5 + 3/6 + 2/6 = 5 + 5/6.

Convert 5 + 5/6 to an improper fraction: 5 + 5/6 = 30/6 + 5/6 = 35/6.

Divide 35/6 by 5 to get 7/6. (Note that 5 = 5/1 such that 35/6 ÷ 5/1 = 35/6 × 1/5 = 35/30 = 7/6. To divide by a fraction, multiply by its reciprocal. The reciprocal of 5/1 is 1/5. To reduce 35/30 to 7/6, divide 35 and 30 each by 5.)

Note that 7/6 feet equates to 1 and 1/6 feet. Since 1 foot = 12 inches, this equates to 1 foot and 2 inches. (It also equates to 14 inches.)

170. Final answer: $625 is the initial balance and $676 is the final balance.

Check the answer: The interest after one year is 0.04 × $625 = $25. The balance after 1 year is $625 + $25 = $650. The interest earned in the second year is 0.04 × $650 = $26. The balance after 2 years is $650 + $26 = $676.

Solution: Method 1: We will use symbols to help keep track of the process.

After one year, the interest earned is $I_1 = 0.04\, P_1$, where P_1 is the original balance (also called the principal). The balance after the first year is $P_2 = P_1 + I_1$. The interest earned for the second year is $I_2 = 0.04\, P_2$. The final balance after two years is $A = P_2 + I_2$. The total interest earned after two years is $I_1 + I_2 = \$51$.

Plug $I_1 = 0.04\ P_1$ and $I_2 = 0.04\ P_2$ into $I_1 + I_2 = \$51$ to get $0.04\ P_1 + 0.04\ P_2 = \$51$.

Plug $P_2 = P_1 + I_1$ into $0.04\ P_1 + 0.04\ P_2 = \$51$ to get $0.04\ P_1 + 0.04\ (P_1 + I_1) = \$51$.

Apply the distributive property: $0.04\ P_1 + 0.04\ P_1 + 0.04\ I_1 = \$51$.

Combine like terms: $0.08\ P_1 + 0.04\ I_1 = \$51$.

Plug $I_1 = 0.04\ P_1$ into $0.08\ P_1 + 0.04\ I_1 = \$51$ to get $0.08\ P_1 + 0.0016\ P_1 = \$51$.

Combine like terms to get $0.0816\ P_1 = \$51$.

Note that $0.0816 = 51/625$. The previous equation is $51\ P_1 / 625 = \$51$.

Multiply both sides by 625 and divide by 51 to get $P_1 = \$625$. This is the initial balance (also called the principal). The final balance after two years is equal to $A = P_1 + I_1 + I_2 = \$625 + \$51 = \$676$. This is the final balance after two years.

Method 2: $A = P_1\ 1.04^2 = 1.0816\ P_1 = P_1 + \$51$ such that $0.0816\ P_1 = \$51$. Divide both sides by 0.0816 to get $P_1 = \$625$. Finally, $A = P_1 + \$51 = \$676$.

171. Final answer: 96% is how much the area of the square is increased.

Solution: Method 1: Increasing the edge length by 40% equates to multiplying the edge length by a factor of 1.4. (This is explained in the note to the solution to Problem 111.) Each edge length is multiplied by 1.4. Since the area of a square equals its edge length times its edge length, the area of the square increases by a factor of $1.4 \times 1.4 = 1.96$. Since the area is multiplied by a factor of 1.96, this means it is increased by 96%. (In this last step, we did the same thing that we did in the first sentence in reverse. See the "Try it with numbers" note below.)

Method 2: We will do the same thing as Method 1, expressed using algebra. Let L be the initial edge length and Y be the final edge length. Then $Y = 1.4\ L$. The initial area is $A = L^2$ and the final area is $B = Y^2 = (1.4L)^2 = 1.4^2\ L^2 = 1.96\ L^2$ using the rule $(cx)^n = c^n x^n$. The ratio of the final area, B, to the initial area, A, is $B/A = 1.96$, which is a 96% increase (since $1.96 = 196\% = 100\% + 96\%$).

Try it with numbers: Suppose that the initial edge length is 10 units. The initial area is therefore $10 \times 10 = 100$ square units. Forty percent of the edge length is $0.4 \times 10 = 4$ units. Since the edge length increased by 40%, the new edge length is $10 + 4 = 14$ units. (Note that this is consistent with multiplying the original edge length by 1.4.) The final area is $14 \times 14 = 196$ square units. Since the initial area was 100 square units, the area increased by $196 - 100 = 96$ square units. This is a 96% increase. (Note that this is consistent with multiplying 100 by a factor of 1.96. An increase of 96% equates to multiplying by 1.96.)

Although making up a number for the length of the square helps to check the answer with numbers, it is not a desirable way to solve the problem. A teacher would likely deduct points for solving a problem by making up numbers.

172. Final answers: A large brick weighs 16 pounds, a medium brick weighs 8 pounds, and a small brick weighs 4 pounds.

Check the answers: 4 pounds is one-half of 8 pounds, and 8 pounds is one-half of 16 pounds. The combined weight is $4 + 8 + 16 = 28$ pounds.

Notes: One way to solve the problem is to think of a medium brick as 0.5 large bricks and a small brick as 0.25 large bricks. The answer is $28 \div 1.75$. Write 1.75 as 7/4 to do this without a calculator.

173. Final answer: 270 of the times displayed on the clock have at least one 2.

Solution: There are 60 times from 12:00 noon thru 12:59 p.m.

There are 60 times from 2:00 p.m. thru 2:59 p.m.

There are 100 times in the 8, 9, 10, 11, 1, 3, 4, 5, 6, and 7 o'clock hours where the minutes are in the twenties, such as 8:20 a.m. thru 8:29 a.m. (Recall that we already counted all of the minutes in the 12 and 2 o'clock hours.)

There are 50 times in the 8, 9, 10, 11, 1, 3, 4, 5, 6, and 7 o'clock hours where the last minute is a 2 and where the minutes are not in the twenties, such as 8:02, 8:12, 8:32, 8:42, and 8:52. (Recall that we already counted all of the times with minutes in the twenties, like 8:22 and 9:22.)

The answer is $60 + 60 + 100 + 50 = 270$.

174. Final answers: 12 lightbulbs are on a short strand, 16 lightbulbs are on a medium strand, and 22 lightbulbs are on a long strand.

Check the answers: $22 \times 15 + 16 \times 10 + 12 \times 8 = 330 + 160 + 96 = 586$

Solution: One way to solve the problem is to use algebra. Let S, M, and L be the number of lights on a small, medium, or long strand, respectively.

$$M = S + 4$$

$$L = M + 6 = S + 10$$

$$8S + 10M + 15L = 586$$

$$8S + 10(S + 4) + 15(S + 10) = 586$$

$$8S + 10S + 40 + 15S + 150 = 586$$

$$33S = 396$$

$$S = 396 \div 33 = 12$$

$$M = S + 4 = 12 + 4 = 16$$

$$L = M + 6 = 16 + 6 = 22$$

175. Final answers: 32 m/s was her first speed and 40 m/s was her second speed.

Note: Recall the three forms of the rate equation from the solution to Problem 10:

$$\text{speed} = \text{distance} \div \text{time}$$

$$\text{distance} = \text{speed} \times \text{time}$$

$$\text{time} = \text{distance} \div \text{speed}$$

Check the answers: Note that 40 m/s is 8 m/s faster than 32 m/s.

Also note that $40 = 1.25 \times 32$. Haley's second speed is 1.25 times faster than her first speed. This is consistent with her time decreasing by a factor of 1.25. To see this, consider the second rate equation above: distance = speed × time. The distance is the same for each attempt. Since Haley's time decreases by a factor of 1.25, her speed must increase by a factor of 1.25 to compensate. (This is an efficient way to solve this problem.) Note that in each case the distance is the same: $40 \times 60 = 2400$ m agrees with $32 \times 75 = 2400$ m.

176. Final answers: The number could be ±1.5 (equivalent to ±3/2).

Check the answers: First check positive 1.5. Eight times the number is 8 × 1.5 = 12. Eighteen divided by the number is 18 ÷ 1.5 = 12.

Now check negative 1.5. Eight times the number is 8 × (–1.5) = –12. Eighteen divided by the number is 18 ÷ (–1.5) = –12.

Solution: One way to solve this problem is to apply algebra. (However, it is not necessary to use algebra to solve the problem.)

$$8x = 18/x$$

$$x^2 = 2.25$$

$$x = 1.5 \text{ or } -1.5$$

In the first step, we multiplied both sides by x and divided both sides by 8. In the last step, we took the square root of both sides. When we take the square root, there are two possible answers because $1.5^2 = 2.25$ and $(-1.5)^2 = 2.25$ also.

177. Final answer: 60 cubes measuring 2 in. × 2 in. × 2 in. can be formed into a rectangular box measuring 6 in. × 8 in. × 10 in.

Solution: Method 1: The first dimension (6 in.) is 3 times wider than 2 in. The second dimension (8 in.) is 4 times wider than 2 in. The last dimension (10 in.) is 5 times wider than 2 in. The answer is 3 × 4 × 5 = 60, as shown below.

Method 2: The volume of the rectangular box is 6 × 8 × 10 = 480 cubic inches. The volume of one small cube is 2 × 2 × 2 = 8 cubic inches. The answer equals 480 ÷ 8 = 60.

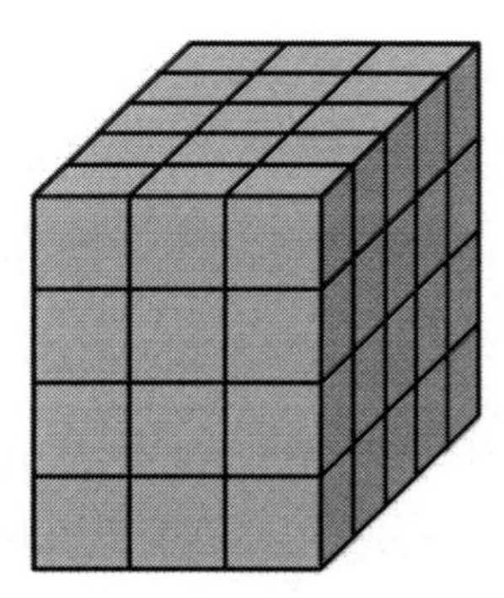

178. Final answer: 12:5 is the ratio of unsold laptops to unsold desktops.
Solution: The ratio of laptops to desktops is initially 2:1.
3/5 of the laptops were sold. This means that 2/5 of the laptops were unsold.
2/3 of the desktops were sold. This means that 1/3 of the desktops were unsold.
In the ratio 2:1, take 2/5 of 2 and 1/3 of 1 to get (2×2/5):(1/3) = (4/5):(1/3) = 12/5.
Notes: The ratio (4/5):(1/3) is equivalent to dividing (4/5) ÷ (1/3). To divide by a fraction, multiply by its reciprocal: (4/5) ÷ (1/3) = (4/5) × (3/1) = 12/5. The ratio 12:5 is equivalent to 2.4:1.
Try it with numbers: Suppose that there are initially 90 laptops. Since there are initially half as many desktops, there are initially 45 desktops. Three-fifths of 90 laptops is 54. Since 54 laptops are sold, 90 – 54 = 36 laptops are unsold. Two-thirds of 45 desktops is 30. Since 30 desktops are sold, 45 – 30 = 15 desktops are unsold. The ratio of unsold laptops to unsold desktops is 36:15, which reduces to 12:5 if you divide 36 and 15 each by 3.
Although making up a number for the number of laptops helps to check the answer with numbers, it is not a desirable way to solve the problem. A teacher would likely deduct points for solving a problem by making up numbers.
179. Final answer: 7.5% is the (average) interest rate earned from the stocks.
Check the answer: 0.075 × $2000 + 0.03 × $5000 = $150 + $150 = $300
Solution: 3% of $5000 equals 0.03 × $5000 = $150. Since the combined interest after one year is $300, the interest earned from the stocks is $300 – $150 = $150. The interest rate for the stocks is $150 ÷ $2000 × 100% = 7.5%.
180. Final answer: 12 is the number.
Check the answer: One more than the number is 12 + 1 = 13. Square this to get 13 × 13 = 169. This agrees with 25 + 12 × 12 = 25 + 144 = 169.
Solution: One way (but not the only way) to solve this problem is with algebra:

$$(n + 1)^2 = 25 + n^2$$

$$n^2 + 2n + 1 = 25 + n^2$$

$$2n = 24$$

$$n = 24/2 = 12$$

In the first step, we applied the f.o.i.l. method: $(x + y)^2 = x^2 + 2xy + y^2$.

181. Final answer: 16 minutes is how much time it takes for the pond to drain.

Solution: 1/40 of the pond drains each minute through the left drain, 1/48 of the pond drains each minute through the middle drain, and 1/60 of the pond drains each minute through the right drain. Add these fractions together to see how much of the pond drains each minute:

$$1/40 + 1/48 + 1/60 = 6/240 + 5/240 + 4/240 = 15/240 = 1/16$$

Note that 15/240 reduces to 1/16 if you divide 15 and 240 each by 15. Since 1/16 of the pond drains each minute, the pond will drain completely in 16 minutes.

Note: It should make sense that it takes much less time for the pond to drain when all three drains are unplugged than if only one drain is unplugged.

182. Final answer: 20 meters is how far the scooter can travel in one second.

Solution: 72 km = 72,000 m and 1 hr. = 3600 s. Combine these to get:

$$72 \text{ km} / 1 \text{ hr.} = 72{,}000 \text{ m} / 3600 \text{ s} = 20 \text{ m/s}$$

183. Final answers: 27:512 is the ratio of the volumes and 9:64 is the ratio of the surface areas.

Notes: Since the volume of a cube equals L^3, each edge length will be cubed to find the volumes. The ratio of the volumes is thus $3^3:8^3 = 27:512$. Similarly, the surface area of a cube is $6L^2$ (since a cube is bounded by 6 square faces). The ratio of the surface areas is $(6\times3^2):(6\times8^2) = 54:384 = 9:64$.

Try it with numbers: Let the edge length of the smaller cube be 3 units. The edge length of the larger cube is then 8 units. The volumes are $3^3 = 27$ and $8^3 = 512$ in cubic units. The ratio of the volumes is 27:512. The surface areas are $6 \times 3^2 = 6 \times 9 = 54$ and $6 \times 8^2 = 6 \times 64 = 384$ in square units. The ratio of the surface areas is 54:384, which reduces to 9:64 if you divide 54 and 384 each by 6. Although making up a number for the edge length helps to check the answer with numbers, it is not a desirable way to solve the problem. A teacher would likely deduct points for solving a problem by making up numbers.

184. Final answer: 72 is the lowest score that Yvonne can earn on one of the next two tests and still be able to have an average score of 90.

Solution: To find the average score, add up the 4 scores and divide by 4. To have an average score of 90, the sum of the 4 scores must equal 360. Subtract the two known scores to find that Yvonne needs to earn at least $360 - 96 - 92 = 172$ points on the next two tests combined. If Yvonne earned a perfect score of 100 on one of the two tests, the other score could be as low as $172 - 100 = 72$.

185. Final answers: The number could be 42 or –42. (The answer is ±42.)

Check the answers: First check the positive answer: $126 \div 42 = 3$ agrees with $42 \div 14 = 3$. Now check the negative answer: $126 \div (-42) = -3$ and $-42 \div 14 = -3$.

Solution: One way (but not the only way) to solve this problem is to apply algebra:

$$126 \div N = N \div 14$$

$$126 \times 14 = N^2$$

$$1764 = N^2$$

$$42 \text{ or } -42 = N$$

In the first step, we cross multiplied. (Multiply both sides of the equation by N and also by 14.) In the last step, we took the square root of both sides. When we take the square root, there are two possible answers because $42^2 = 1764$ and $(-42)^2 = 1764$ also.

186. Final answer: 12 trees are in the group.

Check the answer: This problem is similar to the handshake formula from the solution to Problem 14, which corresponds to a complete graph:

$$12 \times 11 \div 2 = 132 \div 2 = 66$$

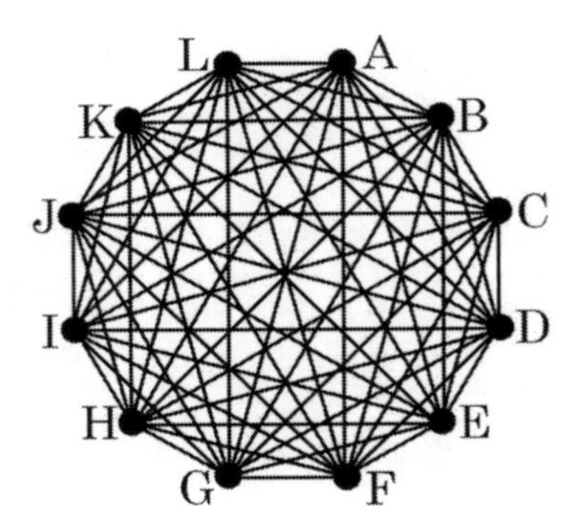

187. Final answer: 17 out of 72 is the probability that at least two of the people were born in the same month.

Solution: Let's call the people A, B, and C.

The probability that B's birth month is different from A's birth month is 11/12. If A and B have different birth months, the probability that C's birth month will be different from the birth months of both A and B is 10/12, which reduces to 5/6. This is called conditional probability. Since this is a conditional probability, to find the probability that A, B, and C all have different birth months, multiply the two probabilities together:

$$(11/12) \times (5/6) = 55/72$$

So far, what we found is the probability that no two of the three people share the same birth month. To find the probability that at least two of the three people share the same birth month, subtract this from one. (Why? The probability that at least two people share the same birth month plus the probability that no two of the three people share the same birth month must equal 100%. If you convert 100% to decimal, you get the whole number 1.)

$$1 - 55/72 = 72/72 - 55/72 = 17/72$$

Note: This is a simple case of a famous probability question called the "birthday problem." If 23 random people are put together (and if it is equally likely for a person to be born on any day of the year), there is approximately a 50% chance that at least one pair will have been born on the same day of the year.

188. Final answer: 108 rotations are completed during this time.

Notes: The key is to realize that when the cylinder completes one rotation, the center of mass of the cylinder travels a distance equal to the circumference (as a result of rolling without slipping). This idea is central to a variety of rolling problems. Convert 3 minutes to 180 seconds. The same rate equation from the solutions to Problems 10, 31, 117, 144, 155, and 164 applies to the center of mass of the cylinder (if you put the time in seconds to be consistent with 1.5 ft/s):

$$\text{distance} = \text{speed} \times \text{time} = 1.5 \times 180 = 270 \text{ feet}$$

Divide the distance traveled by the circumference to determine the number of rotations: $270 \div 2.5 = 108$ rotations.

189. Final answer: Patrick is 21 years old.

Check the answer: $1/28 + 1/21 = 3/84 + 4/84 = 7/84 = 1/12$.

Notes: One way to solve this problem is to apply algebra (but it is not the only way to solve this problem). For example, beginning with $1/28 + 1/P = 1/12$, first find $1/P = 1/12 - 1/28 = 7/84 - 3/84 = 4/84 = 1/21$ and then take the reciprocal of both sides to get $P = 21$. Note that it is incorrect to take the reciprocal of each term of $1/28 + 1/P = 1/12$. If you take the reciprocal of both sides of $1/28 + 1/P = 1/12$, what you get is the nested expression $1/(1/28 + 1/P) = 1/12$. Since the left side does not simplify nicely, it is better to wait until you isolate $1/P$ before you take the reciprocal of both sides. There is, however, a simple way to make the algebra simpler. Multiply both sides of $1/28 + 1/P = 1/12$ by $84P$ to remove all of the denominators. This gives you $3P + 84 = 7P$, for which $84 = 4P$ and $21 = P$. (It pays to remember this "magic" – or "math-ic" trick.)

190. Final answer: 5% is the interest rate.

Check the answers: After one year, $0.05 \times \$500 = \$25$ interest is earned, making the new balance \$525. After two years, $0.05 \times \$525 = \$26.25$ additional interest is earned, making the final balance $\$525 + \$26.25 = \$551.25$.

Solution: Recall the equations from the solution to Problem 170, especially for Method 2. In this problem, $A = P_1 (1 + r)^2$ becomes $\$551.25 = \$500 (1 + r)^2$. Divide by \$500 on both sides: $1.1025 = (1 + r)^2$. Note that 1.1025 = 441/400. Square root both sides of the equation to get 21/20 = 1 + r. In decimal form, this is 1.05 = 1 + r. Subtract 1 from both sides: 0.05 = r. The interest rate is 5%.

191. Final answers: Sally has 144 small squares and Tom has 36 small squares. Check the answers: 144 + 36 = 180. Sally arranged her 144 small squares in the shape of a 12 × 12 square. Tom arranged his 36 small squares in the shape of a 6 × 6 square. Note that 12 = 2 × 6.

Solution: One way (but not the only way) to solve this problem is with algebra:

$$S^2 + T^2 = 180$$

$$S = 2T$$

$$(2T)^2 + T^2 = 180$$

$$4T^2 + T^2 = 180$$

$$5T^2 = 180$$

$$T^2 = 36$$

$$T = 6$$

$$S = 2T = 2 \times 6 = 12$$

192. Final answer: 5:00.

Solution 1: From 12:00 to 4:00 "should" be 4 hours = 240 minutes, but more time has actually passed because the clock is running slow. Every 24 of the clock's minutes correspond to 30 real minutes. Since 240 of the clock's minutes pass from 12:00 to 4:00, this means that 300 real minutes have passed. Since 300 real minutes = 5 real hours, the real time is 5:00.

Solution 2: Since 30:24 reduces to 5:4, real time is passing 5/4 = 1.25 times faster than the clock's time. Multiply the clock's 4 hours of passage by 1.25 to find that 4 × 1.25 = 5 real hours have passed.

193. Final answer: 511 m will be the height when the width is 8 m.

Notes: Raise the width to the third power and subtract one to find the height:

$$2^3 - 1 = 8 - 1 = 7 \text{ m}$$

$$3^3 - 1 = 27 - 1 = 26 \text{ m}$$

$$4^3 - 1 = 64 - 1 = 63 \text{ m}$$

$$5^3 - 1 = 125 - 1 = 124 \text{ m}$$

$$8^3 - 1 = 512 - 1 = 511 \text{ m}$$

It is not necessary to know any special formulas to answer this question. You just need to be able to identify the pattern.

194. Final answer: 7:5 is the ratio of the perimeters of the two rectangles.

Check the answers: As shown below, in order for the ratio of the areas to equal three, one rectangle must be three times as wide as the other. This means that the widths of the two rectangles are 18 cm and 6 cm. The perimeter of the wider rectangle is $2 \times 24 + 2 \times 18 = 48 + 36 = 84$ cm. The perimeter of the narrower rectangle is $2 \times 24 + 2 \times 6 = 48 + 12 = 60$ cm. The ratio of the perimeters equals 84:60 = 7:5. (Note that 84:60 reduces to 7:5 if you divide 84 and 60 each by 12.)

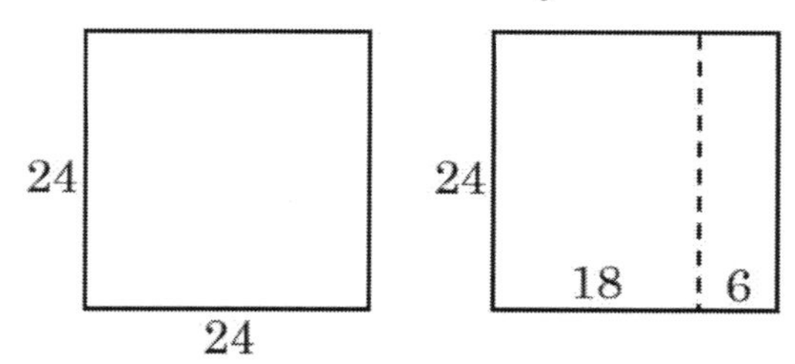

195. Final answers: The average speed must be less than $(20 + 30) \div 2 = 25$ m/s because the robot spends more time traveling 20 m/s and less time traveling 30 m/s. The actual average speed is 24 m/s.

Notes: In math class, we often find the average of two numbers by adding them together and dividing by two, but that does not work for this problem. That kind of average is called the arithmetic mean. If you found the arithmetic mean of

20 m/s and 30 m/s, you would get (20 + 30) ÷ 2 = 25 m/s. The arithmetic mean only equals the average speed when all of the times are equal. In this problem, the times are not equal, so the arithmetic mean is **not** the average speed.

To better understand why that is, consider the following extreme case. Suppose that the robot traveled 30 m/s for exactly 1 second and then traveled 20 m/s for a total of 1,000,000 hours. If the robot spent almost the entire time traveling 20 m/s and only 1 second traveling 30 m/s, it would not make sense to say that the robot's average speed is 25 m/s. Clearly, the average speed should be very close to 20 m/s in such an extreme case.

Now we will return to the actual problem (not the extreme case). The main idea is that the time spent traveling each speed matters. The robot will spend more time traveling 20 m/s and less time traveling 30 m/s. The average speed will therefore be closer to 20 m/s than it is to 30 m/s, which means that it must be less than 25 m/s.

Solution: Method 1: Use the definition of average speed stated in the problem.

$$\text{average speed} = \text{total distance} \div \text{total time}$$

Let the distance for each trip be D, so that the total distance traveled is 2D. The total time will be $t_1 + t_2$ (the sum of the times for each trip).

$$\text{average speed} = 2D \div (t_1 + t_2)$$

Recall the rate equation from the solutions to Problems 10, 31, 117, 144, 155, 164, and 175:

$$\text{distance} = \text{speed} \times \text{time}$$

Apply this to each trip:

$$D = 30\ t_1 \text{ and } D = 20\ t_2$$

Divide both sides by 30 in the left equation and by 20 in the right equation:

$$D/30 = t_1 \text{ and } D/20 = t_2$$

Substitute these expressions for time into the equation for average speed.

average speed = 2D ÷ (D/30 + D/20)

To add D/30 + D/20, make a common denominator: D/30 + D/20 = 2D/60 + 3D/60 = 5D/60 = D/12. Plug this into the equation for average speed.

average speed = 2D ÷ (D/12)

To divide by a fraction, multiply by its reciprocal. The reciprocal of D/12 is 12/D.

average speed = 2D × 12/D

average speed = 2 × 12 = 24 m/s

Method 2: Add the speeds in reciprocal, and divide 2 by that sum:

2 / (1/30 + 1/20) = 2 / (2/60 + 3/60) = 2 / (5/60) = 2 / (1/12) = 2 × 12/1 = 24

The second method basically condenses the work from the first method down to one formula with a few simple steps.

Note: Average speed questions occasionally show up in math competitions.

196. Final answers: 2A × B2 = 2778 with all three numbers written in base 12. If all three numbers are written in base 10, it is 34 × 134 = 4556.

Notes: The first 36 numbers are written in base 12 and base 10 below.

Base 12	1	2	3	4	5	6	7	8	9	A	B	10
Base 10	1	2	3	4	5	6	7	8	9	10	11	12

Base 12	11	12	13	14	15	16	17	18	19	1A	1B	20
Base 10	13	14	15	16	17	18	19	20	21	22	23	24

Base 12	21	22	23	24	25	26	27	28	29	2A	2B	30
Base 10	25	26	27	28	29	30	31	32	33	34	35	36

These tables show that 2A in base 12 is equivalent to 34 in base 10.

Observe that 144 in base 10 is equivalent to 100 in base 12. If we count backwards from these numbers, we can figure out what B2 means in base 12:

Base 12	B1	B2	B3	B4	B5	B6	B7	B8	B9	BA	BB	100
Base 10	133	134	135	136	137	138	139	140	141	142	143	144

The previous table shows that B2 in base 12 is equivalent to 134 in base 10. We thus see that 2A × B2 in base 12 is equivalent to 34 × 134 in base 10. In base 10, it is easy to determine that 34 × 134 = 4556. Now we need to convert 4556 from base 10 to base 12 to determine the answer to 2A × B2 in base 12. Note the following helpful conversions between base 10 and base 12:

Base 10	$12^1 = 12$	$12^2 = 144$	$12^3 = 1728$	$12^4 = 20,736$
Base 12	10	100	1000	10,000

In base 10, we can rewrite 4556 as 4556 = 2 × 1728 + 7 × 144 + 7 × 12 + 8. This shows that 4556 in base 10 is equivalent to 2778 in base 12. (You may wish to review the solution to Problem 123 where we did a similar conversion to base 3.)

197. Final answer: 35% of all of the rocks are shiny.

Try it with numbers: Suppose that the first bag has 100 rocks. Since 25% of the rocks in the first bag are shiny, the first bag will have 25 shiny rocks and 75 rocks that are not shiny. The second bag will then have 200 rocks. Since 40% of the rocks in the second bag are shiny, the second bag will have 0.4 × 200 = 80 shiny rocks and 200 – 80 = 120 rocks that are not shiny. When the two bags are mixed, there will be 25 + 80 = 105 shiny rocks and a total of 300 rocks. Divide 105 by 300 to get 0.35, and multiply by 100% to convert 0.35 to a percent: 0.35 = 35%. Although making up a number for the number of rocks in the first bag helps to check the answer with numbers, it is not a desirable way to solve the problem. A teacher would likely deduct points for solving a problem by making up numbers. One way to solve this problem is to call the number of rocks in the first bag N. The number of shiny rocks in the first bag is then 0.25N. The number of rocks in the second bag is 2N and the number of shiny rocks in the second bag is 0.4 × 2N = 0.8N. The total number of shiny rocks is 0.25N + 0.8N = 1.05 N and the total number of rocks is N + 2N = 3N. The fraction of shiny rocks is 1.05/3, which reduces to 0.35. Convert this to a percent: 0.35 = 35%.

198. Final answer: 2 out of 5 is the probability that the seating chart will have both girls seated next to one another.

Solution: Label the six chairs A, B, C, D, E, and F. There are 6 different ways that the two girls could be seated next to one another:

A and B, B and C, C and D, D and E, E and F, F and A

Since the table is round and the chairs are equally spaced, note that chair F is next to chair A. There are 9 different ways that the two girls could be seated so that they are not next to one another:

A and C, A and D, A and E, B and D, B and E,
B and F, C and E, C and F, D and F

The total number of ways that the girls can be seated is 6 + 9 = 15. The likelihood that the two girls will be seated next to one another is 6/15, which reduces to 2/5 (if you divide 6 and 15 each by 3).

199. Final answer: 80 minutes is how much time it takes.

Solution: The father empties 50/60 = 5/6 gallons of water each minute. The son empties 75/60 = 5/4 gallons of water each minute. (Note that 75 and 60 are each evenly divisible by 15.)

$$5/6 + 5/4 = 10/12 + 15/12 = 25/12$$

Each minute, the father and son remove 25/12 gallons of water. However, some water is entering the boat through a leak in the bottom. The leak causes 20/60 = 1/3 gallons of water to enter through the bottom of the boat each minute. To find the net amount of water that the father and son remove from the boat each minute, subtract the fraction for the leak:

$$25/12 - 1/3 = 25/12 - 4/12 = 21/12 = 7/4$$

Note that 21/12 reduces to 7/4 if you divide 21 and 12 each by 3. Divide 140 by 7/4 to determine how much time it will take to remove a net 140 gallons from the boat. To divide by a fraction, multiply by its reciprocal.

$$140 \div (7/4) = 140 \times (4/7) = 560/7 = 80$$

It takes 80 minutes for the father and son to temporarily empty water from the boat. (Hopefully, they will now repair the leak before more water comes in.)

Notes: 80 minutes is equivalent to 80/60 = 4/3 hours (or 1 hour and 20 minutes). Think about it: The father and son empty 50 + 75 gallons of water each hour. The boat has 140 gallons of water in it initially. If the boat did not have a leak, it would take 140 ÷ 125 = 1.12 hours = 1 hour + 7.2 minutes to empty the boat. Due to the leak, the father and son need to empty an additional 20 gallons of water each hour. As a result, it takes 80 minutes = 1 hour + 20 minutes to empty the boat. It took about 13 minutes longer for the father and son to empty the boat due to the leak.

200. Final answer: 375 meters is the total distance that the squirrel travels.

Note: Recall the three forms of the rate equation from the solution to Problem 10:

$$\text{speed} = \text{distance} \div \text{time}$$

$$\text{distance} = \text{speed} \times \text{time}$$

$$\text{time} = \text{distance} \div \text{speed}$$

Solution: Use the third form of the rate equation to determine how much time it takes for the two dogs to meet. Since each dog travels 4 m/s, the two dogs get 8 m closer to each other every second. Divide 200 m by 8 m/s to determine when the two dogs meet.

$$\text{time} = 200 \div 8 = 25 \text{ seconds}$$

Use the middle equation to determine the total distance that the squirrel travels.

$$\text{distance} = 15 \times 25 = 375 \text{ meters}$$

Note: The squirrel will be caught when the two dogs meet one another. We used this to simplify the solution. It would be unnecessarily tedious to break the trip up into several pieces, solving for how far the squirrel travels until reaching the first dog, how far the squirrel travels until reaching another dog, etc.

TOPIC LIST

WAS THIS BOOK HELPFUL?

Much effort and thought were put into this book, such as:

- Careful selection of problems for their instructional value.
- Providing explanations and helpful ideas in the answer key.
- Including problem-solving tips and examples in the Idea Center.

If you appreciate the effort that went into making this book possible, there is a simple way that you could show it:

Please take a moment to post an honest review.

For example, you can review this book at Amazon.com or Goodreads.com. Even a short review can be helpful and will be much appreciated. If you are not sure what to write, following are a few ideas, though it's best to describe what is important to you.

- Did you enjoy the selection of problems?
- Which problems did you like? What did you like about them?
- Were the explanations in the answer key clear?
- Did you learn anything from the answer key?
- Was the answer key helpful?
- Was the Idea Center helpful?
- Would you recommend this book to others? If so, why?

Do you believe that you found a mistake? Please email the author, Chris McMullen, at greekphysics@yahoo.com to ask about it. One of two things will happen:

- You might discover that it wasn't a mistake after all and learn why.
- You might be right, in which case the author will be grateful and future readers will benefit from the correction. Everyone is human.

ABOUT THE AUTHOR

Dr. Chris McMullen has over twenty years of experience teaching university physics in California, Oklahoma, Pennsylvania, and Louisiana. Dr. McMullen is also an author of math and science workbooks. Whether in the classroom or as a writer, Dr. McMullen loves sharing knowledge and the art of motivating and engaging students.

The author earned his Ph.D. in phenomenological high-energy physics from Oklahoma State University in 2002. Originally from California, Chris McMullen earned his Master's degree from California State University, Northridge, where his thesis was in the field of electron spin resonance.

As a physics teacher, Dr. McMullen observed that many students lack fluency in fundamental math skills. In an effort to help students of all ages and levels master basic math skills, he published a series of math books on arithmetic, fractions, long division, algebra, geometry, trigonometry, logarithms, and calculus entitled *Improve Your Math Fluency*. Dr. McMullen has also published a variety of science books, including astronomy, chemistry, and physics workbooks.

MATH

This series of math workbooks is geared toward practicing essential skills:

- Prealgebra and algebra
- Geometry and trigonometry
- Logarithms and exponentials
- Calculus
- Fractions, decimals, and percentages
- Long division
- Multiplication and division
- Addition and subtraction
- Roman numerals
- The four-color theorem and basic graph theory

www.improveyourmathfluency.com

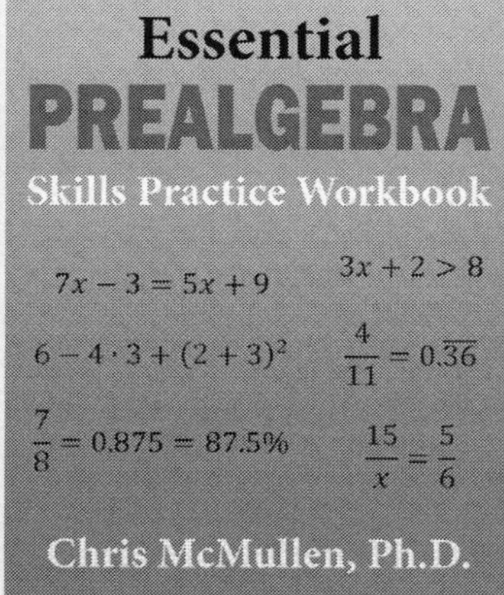

PUZZLES

The author of this book, Chris McMullen, enjoys solving puzzles. His favorite puzzle is Kakuro (which is kind of like a cross between crossword puzzles and Sudoku). He once taught a three-week summer course on puzzles. If you enjoy mathematical pattern puzzles, you might appreciate:

300+ Mathematical Pattern Puzzles

Number Pattern Recognition & Reasoning

- Pattern recognition
- Visual discrimination
- Analytical skills
- Logic and reasoning
- Analogies
- Mathematics

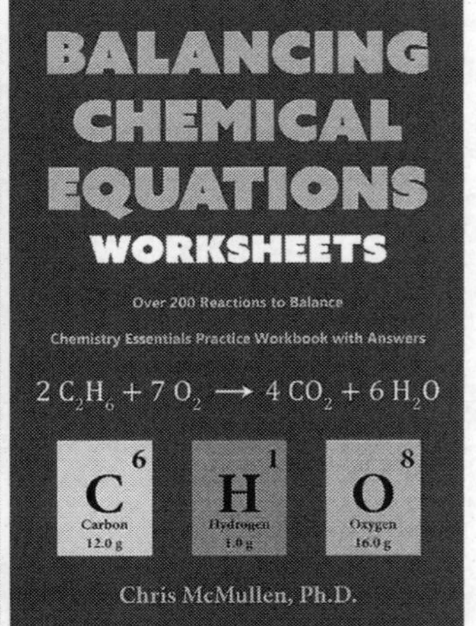

THE FOURTH DIMENSION

Are you curious about a possible fourth dimension of space?

- Explore the world of hypercubes and hyperspheres.
- Imagine living in a two-dimensional world.
- Try to understand the fourth dimension by analogy.
- Several illustrations help to try to visualize a fourth dimension of space.
- Investigate hypercube patterns.
- What would it be like to be a 4D being living in a 4D world?
- Learn about the physics of a possible four-dimensional universe.

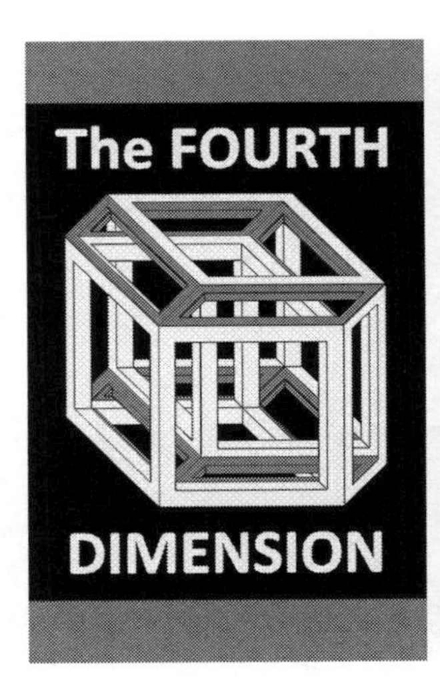

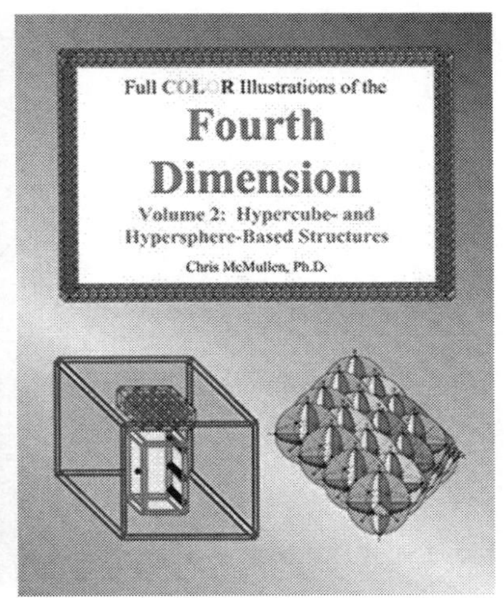

SCIENCE

Dr. McMullen has published a variety of **science** books, including:

- Basic astronomy concepts
- Basic chemistry concepts
- Balancing chemical reactions
- Calculus-based physics textbooks
- Calculus-based physics workbooks
- Calculus-based physics examples
- Trig-based physics workbooks
- Trig-based physics examples
- Creative physics problems
- Modern physics

www.monkeyphysicsblog.wordpress.com

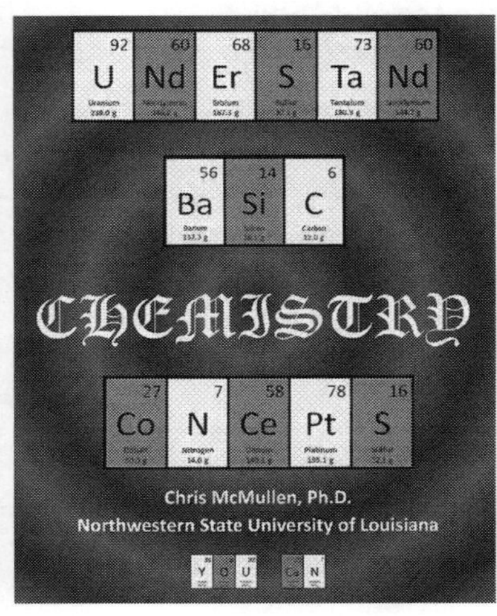

Made in the USA
Middletown, DE
18 February 2023

25135598R00110